X-RAY REPAIR

ABOUT THE AUTHOR

Joseph J. Panichello earned his B.S. degree in Biology from Pfeiffer University, an A.A.S. degree in Biomedical Equipment Technology from Stanly Technical College, and is AAMI certified. He has worked as an in-house BMET, an x-ray field service engineer, and as a service manager for an independent x-ray service company. He lives in Medford, New Jersey with his wife and daughter.

Second Edition

X-RAY REPAIR

A Comprehensive Guide to the Installation and Servicing of Radiographic Equipment

By

JOSEPH J. PANICHELLO, C.B.E.T.

CHARLES C THOMAS • PUBLISHER, LTD.
Springfield • Illinois • U.S.A.

Published and Distributed Throughout the World by

CHARLES C THOMAS • PUBLISHER, LTD.
2600 South First Street
Springfield, Illinois 62704

ISBN 0-398-07537-9 (hard)
ISBN 0-398-07538-7 (paper)

Library of Congress Catalog Card Number: 2004051702

With THOMAS BOOKS *careful attention is given to all details of manufacturing and design. It is the Publisher's desire to present books that are satisfactory as to their physical qualities and artistic possibilities and appropriate for their particular use.* THOMAS BOOKS *will be true to those laws of quality that assure a good name and good will.*

Printed in the United States of America
SR-R-3

Library of Congress Cataloging-in-Publication Data

Panichello, Joseph J.
X-ray repair : a comprehensive guide to the installation and servicing of radiographic equipment / by Joseph J. Panichello.-- 2nd ed.
p. cm.
Includes bibliographical references and index.
ISBN 0-398-07537-9 -- ISBN 0-398-07538-7 (pbk.)
1. Radiography, Medical--Equipment and supplies--Maintenance and repair. I. Title.

RC78.5.P26 2004
681′.761--dc22

2004051702

To Caroline

SPECIAL DEDICATION

To Mr. M. Maung (1941–1996)

This book is in part dedicated to the life and memory of Mr. Maung Maung, internationally certified clinical engineer of Newcomb Medical Center in Vineland, New Jersey. Maung first introduced me to medical instrumentation, helped me to get accepted into the school for Biomedical Equipment Technology, allowed me to serve my internship under him, and helped write the resume (and initiated the phone call!) that led to my first job in this field.

Throughout the years, Maung was my good friend and always provided sound advice in all aspects of life for which I am truly indebted. He even provided some valuable direction when, in confidence, I informed him of my desire to write this book. He was the most giving human being I have ever met.

PREFACE

During the past six years since this book was first published, I have received excellent feedback from a number of readers. Most expressed that they were glad to have a book solely dedicated to their profession: some offered suggestions as to how to make the book more "comprehensive." This second edition was written in response to those comments, as well as to the suggestions made by course instructors and service engineers who are currently employed. Thanks to all for your helpful comments. The additional information provided here, I hope, will help students and engineers to better understand x-ray servicing. My goal is to make *X-Ray Repair* a single source for general x-ray servicing questions.

Since the book was first published, the medical imaging field has continued to grow at an extremely rapid pace. There have been great strides in all aspects of radiography including advances in image acquisition, image processing, and image storage. Moreover, advances in digital technology are the impetus that drive this field today.

Fortunately for service engineers, the basics of x-ray production still apply to the field of radiography and will do so for many years to come. Even though the use of digital technology is becoming commonplace, the means of x-ray production has not changed. In fact, each device currently found in a general radiographic room (i.e., the x-ray generator, tube support, table, and wall unit) is still needed to obtain that quality image.

This new edition includes all of the information found in the first edition and will cover two additional specialized areas of radiography: tomography and mammography. These new chapters provide the information required to allow a service person at the entry level to begin servicing tomography and mammography equipment with a high degree of confidence.

The chapter on tomography presents all of the essentials required to operate and service a tomographic unit. A tomographic unit is, in essence, a standard radiographic unit with the addition of a few specialized components. This chapter focuses on the design and function of those components. Tomographic theory is likewise discussed, and all of the required maintenance and calibration procedures are outlined.

Mammography, another specialized field of radiography, requires a different servicing approach from that used in general radiography. This chapter provides the "basics" of breast imaging and includes a detailed description of a mammographic unit. Electronic calibrations and troubleshooting techniques are also covered. Finally, an overview of digital mammography is included at the end of the chapter.

Readers also suggested that the subject of fluoroscopy be included. I have decided, however, not to include this subject in this edition. Although related to radiography, fluoroscopy requires servicing techniques that are somewhat advanced and, therefore, should be covered in a later edition, or separate book. In this way, I can maintain the central focus of this book, which is to cover all aspects of general radiography.

I have rearranged the order of chapters to accommodate the new material, starting with the most basic concepts and ending with the more advanced specialty fields. Consequently, the new chapters on tomography and mammography can be found at the end of the book.

I have also included an informative chapter entitled "The Seasoned Engineer" which describes specific practices that will allow engineers who are relatively new to the field to advance in their profession. The chapter provides the necessary tools needed for the novice to become proficient in x-ray servicing as quickly as possible, pointing a clear path to becoming an imaging specialist. This final chapter of the book aims to provide inspiration and the proper mindset to approach the challenging field of x-ray servicing.

I would like to mention a few people who have provided technical input for this edition: Tom Avellino Jr., senior field service engineer; Hugh Hudson, technical support engineer (Hologic Inc.); Dennis King, mammography specialist; Fred Lacy, senior applications engineer (Varian Products); and Karl Schmidt, field service engineer.

The diagrams for the new chapters were created by Bob Verstege, graphics designer and friend. I am very grateful for his help.

Thanks, again, to Dr. Vallin D. Estes for his help with this edition and for his continual support and friendship.

Lastly, I want to thank Michael Payne Thomas, editor of Charles C Thomas Publisher, Ltd., for encouraging me to offer this second edition and for his valuable suggestions regarding new material for the book.

It has been a great pleasure working on this project and I am thankful for having the opportunity. Good luck to all students.

Joe Panichello

PREFACE TO FIRST EDITION

This book is intended to be used as a field guide and reference manual for field service engineers and in-house biomedical engineers when servicing radiographic equipment. It is also written in a way so as to prepare the student of x-ray servicing for all the specific duties involved for the safe and proper maintenance of radiographic equipment. The goal is to provide standard servicing practices that, at the present time, are not available.

The idea for this book arose during my career as an x-ray service engineer.[1] While working as a field service engineer, I observed many different approaches to the installation, preventive maintenance, and repair of radiographic equipment. Some of these approaches were more successful than others, but since standards for servicing x-ray equipment were nonexistent, the service engineer would not always know if he or she had made the best choices while performing service. As a result, any two x-ray service engineers may perform preventive maintenance (or PM) on identical radiographic rooms in completely different ways. This is a serious problem that often reflects badly on our profession. When a customer pays for a PM, what service is he or she actually getting?

The background of the x-ray service engineer of today varies greatly, ranging from no formal education in electronics to degreed electrical engineers. X-ray servicing is one of the few remaining technical fields where minimum requirements for education and practical training have not yet been established, although this situation is slowly changing. A typical x-ray service engineer working in the field may

1. The term "x-ray service engineer" has become the accepted title for individuals who install, repair, and maintain radiographic equipment and will be used throughout this book. It encompasses all the commonly used titles such as, x-ray engineer, x-ray service technician, field service engineer, and imaging specialist.

have some electronic background, but most of his or her x-ray servicing skills are developed on the job.

To better understand the wide discrepancy in educational background of the x-ray service engineer, we must look back at the history of x-ray servicing. In the early years, servicing x-ray units mainly required physical strength and good mechanical ability. The reason for this was that x-ray units generally were large and cumbersome, usually requiring several men to install and service.[2] In fact, the physical size of the individual was often a major criterion for hiring in this field.

Not only was physical strength needed to install and repair equipment, but an exceptional mechanical ability was also a requirement. Service engineers had to work proficiently with mechanical parts and complicated gear assemblies. They needed skills for cutting, grinding, drilling, and tapping metal parts in order to make covers, brackets, and supports fit properly into place. Moreover, if a part was needed for a particular repair but could not be purchased, the x-ray service engineer would have to design and make the part himself! Also, since many of the devices in a radiographic room had to be securely mounted to ceilings, walls, and floors, the service engineer required some knowledge of building construction.

X-ray equipment found in the early radiographic rooms predominantly utilized electromechanical devices to perform the various functions. These included large mechanical tap switches, relay contactors, electromagnetic lock assemblies, AC motors, and motor drive systems. Sophisticated electronic components and circuit boards were not widely used in these early machines and, consequently, it was advantageous for the service engineer to mainly possess the technical skills required to properly service electromechanical devices.

Along with the technical requirements, there were also a certain amount of safety risks associated with working in this field. Because of the potentially lethal voltages involved in x-ray circuits, and the nature of radiation itself, this job was definitely not for the fainthearted. These factors would limit the number and type of individual who would enter this field.

As a result, only a small number of skilled x-ray service engineers were available in those days to work in this potentially hazardous occu-

2. Until very recently, x-ray servicing was a male-dominated field.

pation. These men were highly skilled and commanded great respect in the medical field. Also, because they were small in number, the engineers had to be self-reliant and capable of working independently, oftentimes for long hours, until the equipment became operational.

Since there was little formal training available for most engineers at that time, and proper documentation was often sketchy at best, no standard methods of servicing radiographic equipment ever evolved. X-ray servicemen would simply develop their own style and approach to servicing the equipment.

This individualistic approach to servicing was successful for some engineers, but tended to cause many service problems in the long run. Engineers often had different ways of testing and repairing equipment, and would use different calibration accuracy standards. Under these circumstances, there would be no way of actually knowing whether or not a radiographic unit was operating at an optimum performance level. In addition, because there were different methods of installing equipment, two identical x-ray units, installed at separate sites (by different engineers), could have completely different performance characteristics and, may have even operated differently. Consequently, these two units could not be serviced in exactly the same way, which poses many problems for the engineer who was unfamiliar with the particular installation.

This approach to servicing is totally unacceptable in the healthcare industry of today. Standards in servicing must be established so that radiographic equipment found in every hospital and clinic will operate at a consistent level of performance. By standardizing installation procedures, routine preventive maintenance, and calibration procedures, consistent performance levels in equipment can easily be achieved. This book will describe those standard procedures in detail. After reading *X-ray Repair,* the student and entry level engineer will gain a better understanding of what we are trying to accomplish in x-ray servicing and, in turn, improve the overall level of the services that are provided.

The book begins with a brief history of the discovery of x-rays and describes their early use in the medical field. An overview of basic x-ray principles is included, as well as a chapter on the individual components of an x-ray system. A chapter on the safety hazards related to working with radiographic equipment is also included. This chapter is especially important to students and entry-level engineers because of the potentially lethal hazards associated with this field.

Next, a step-by-step guide for the proper installations of radiographic equipment is given. Here, all aspects of an installation are discussed, starting with the initial planning of the room to user inservicing and final room checkout. A chapter on preventive maintenance explains why PM's are performed, and demonstrates the correct procedure for performing preventive maintenance on radiographic equipment.

A troubleshooting chapter provides effective troubleshooting techniques that can be used on any type of radiographic equipment. Once a failure has been identified through these troubleshooting techniques, the proper steps for repairing and testing the x-ray system are given so that there will not be any service recalls on the repair. In addition, specific problems that regularly occur in radiographic rooms are discussed with suggestions included on how to correct them. Finally, a chapter describing ways to establish (and maintain) good customer relations is included.

In order to limit the size of the book, the discussions concentrate on the General Radiographic Room. Specialized areas such as Mammography, Tomography, Special Procedures, Cath Labs, and Computerized Tomography (CT) are not included. Although these special applications of x-ray use the same basic principles as in general radiography, they require different servicing techniques and would best be covered in another text. Mobile x-ray units, however, are discussed because of their widespread use in general radiography.

At the end of the book the service engineer will find convenient charts to make his or her job easier when working out in the field. Also included in the Appendix is a complete list of the tools and test equipment needed to properly service x-ray equipment. The glossary will provide a quick reference for any unfamiliar terms used throughout the text.

It is my hope that this book will greatly aid the x-ray service engineer in servicing radiographic equipment and that it will play a role in improving the overall level of x-ray service that is currently being provided.

Joseph J. Panichello

ACKNOWLEDGMENTS

There are many people who have helped and inspired me while pursuing a career in the medical electronics field. I would like, at this time, to acknowledge those who most influenced me as my career progressed to the x-ray servicing field and, ultimately, to the writing of this book.

First and foremost, I would like to thank the original "Boys in Blue" of Deborah Heart and Lung Center, located in Browns Mills, New Jersey: Larry Melson for demonstrating a standard of excellence in servicing medical equipment; John Ludwig for advising me to pursue a career in the field of x-ray service and for his support over the years; Arch Tannock for demonstrating "the basics" of x-ray servicing; and finally, Ed Rourke for teaching me the finer points of x-ray imaging and demonstrating the rewards that this field has to offer. Ed has been a role model and good friend throughout my career.

I would also like to thank Dr. J. Eldridge of Deborah Heart and Lung Center for selling to me, at a very reasonable price, his notebook computer on which I wrote this book. Thanks Jay!

I would like to thank Mr. Tom Avellino of Avellino X-ray Corporation for sharing much of the knowledge that he accumulated while servicing x-ray equipment over the last thirty years.

To Mr. David Eveland of South Jersey X-ray for helping me to understand x-ray servicing from the business point of view. Thanks Dave!

I would like to thank Mr. Mark DePalma of Atco, New Jersey for creating all of the line drawings that are included in this book. Mark, an artist who normally works in the freehand environment, had his skills (and patience) tested as he had to create these precision drawings. He lived up to the challenge by producing high quality, detailed drawings in a relatively short period of time. Thanks Mark!

I would like to acknowledge the faculty of Stanly Community College for the great job they did in preparing me for a career in medical electronics. In particular, I would like to thank Mr. Doug Weaver, CBET, for his advice and friendship over the years.

I would like to thank Pfeiffer College, Misenheimer, North Carolina, for providing me with a quality education and the tools that I will use throughout my life, and in particular, those members of the science faculty who showed me a different way to view the world.

I would especially like to acknowledge Dr. Vallin D. Estes, Professor of Language and Literature at Pfeiffer College. If not for Dr. Estes, I'm not sure this book could have been written. Dr. Estes, my Latin professor at that time, introduced me to the "classics" of literature and in doing so, demonstrated the beauty of life through art and ultimately provided me with what I term "a real education." He became my mentor while in college and in the last fifteen years, a close friend. Dr. Estes edited the first drafts of this book and continually provided the encouragement and motivation that I desperately needed to complete it. Thanks Dayton–for everything.

I would like to thank Charles C Thomas, Publisher for considering my manuscript. It was a pleasure working with such a prestigious publishing firm.

Finally, I would like to thank my wife, Caroline, for her patience and understanding while I was writing this book. This project proved to be more difficult than I ever imagined and I could not have done it without her. Thanks Caroline!

THE LAWS OF X-RAY SERVICE

I. When radiographic equipment is installed properly, it will perform more reliably and for a longer period of time. When a failure does occur, it is more easily repaired.

II. If PM's are performed properly and at regular intervals, equipment failure rate will be greatly reduced and the equipment will operate well beyond its predicted life expectancy.

III. When servicing equipment, if the problem is correctly diagnosed, properly repaired, and thoroughly tested, there will be no service recalls for that service call.

The service engineer should adhere to these laws at all times when servicing radiographic equipment. Without question, if the engineer performs his or her duties as outlined in this book, equipment reliability and longevity will absolutely increase. In addition, there will be a greater level of customer satisfaction, which is, of course, the ultimate goal.

The actual quality of the particular x-ray equipment can, however, alter any predictions made by the "Laws." If the equipment is of exceptional quality, but the service engineer does not follow the "Laws of X-ray Service," the higher failure rates will result in lower customer satisfaction. If the equipment is just average in performance and price, a good service engineer who follows the "Laws" will get optimum performance out of the room and achieve a high level of customer satisfaction. Cheaply manufactured equipment, regardless of quality of service, will fail often and will eventually be removed from the site!

CONTENTS

FIGURES

X-RAY REPAIR

Chapter I

THE DISCOVERY OF X-RAYS

The use of x-rays is so prevalent in today's society that it is difficult to imagine how many industries, especially the medical profession, could function without their daily use. Indeed, most people have had some contact with x-rays in their lifetime and are aware of their importance. For example, doctors, structural engineers, research scientists, and even airline personnel all use x-rays routinely in their respective careers. Hardly anyone would argue that the discovery of x-rays was one of the most important discoveries of our time. Mankind has certainly benefited from their use.

Because of their widespread use, it is difficult to imagine that x-rays were discovered only a century ago. Grandparents of some of the readers of this book were growing up during the time when x-rays were first being demonstrated. And though there have been many advancements made in radiographic equipment design since that time, most of the devices used today in the general radiographic room very closely resemble those early units. The service engineer can gain a much better understanding of general x-ray theory by learning just how x-rays were first discovered and then following the development of the x-ray tube and radiographic system.

The discovery of x-rays, like many other important discoveries, occurred accidentally while scientists were experimenting with glass vacuum tubes. The vacuum tubes that were used were made of thin-walled glass and contained two metal electrodes. They were partially evacuated by a mercury pump and then filled with a specific gas. When a high tension discharge from an induction coil was applied to the vacuum tube, beautifully-colored streams of light were produced between the two electrodes of the tube. These streams of light mostly were a source of fascination and curiosity for the scientists at that time.

After further experimentation with improved vacuum tube designs, scientists began to learn more about the streams of light. They determined that the stream of light originated at the negative electrode (cathode) within the tube and flowed to the positive electrode (anode). In addition, as the tube's vacuum level was increased, the streams would likewise change in color and character, until they finally became invisible.

These "rays" were labeled ***cathode rays*** by William Crookes in 1878, who was the first to observe their directional properties while experimenting with vacuum tubes.[3] During his experiments performed in the dark, Crookes noticed that the cathode rays appeared outside of the glass tube as a bluish glow. He later modified the glass tubes by placing an aluminum window in the tube so that the cathode rays could pass through more easily. These modified tubes then became known as ***Crookes Tubes*** and were commonly used for experimentation.

The cathode rays could easily pass through the aluminum window of the tube and would produce luminescent effects on phosphor materials. They would also affect photographic plates. These rays, however, were of very low energy and, consequently, would be absorbed in only a few centimeters of air. Unfortunately, Crookes had not realized that what he really was observing were low energy x-rays.

Seventeen years later, scientists were still experimenting with the Crookes Tubes. Philip Lenard demonstrated that the cathode rays scattered in all directions at the point where they exited the aluminum window. He also made shadow pictures termed "sciagraphs" on photographic plates. After many experiments he concluded that there was another unidentified component to the cathode rays.

The person credited with the discovery of "x-rays" was a German physicist by the name of Wilhelm Conrad Roentgen. Roentgen was born in Lennep, Germany, in 1845. By the age of twenty-four, he had already published many scientific papers on the properties of gasses. During his career as professor and scientist, he published more than forty papers on scientific phenomena and was highly respected by his fellow scientists.

---

3. The cathode ray phenomenon is commonly used today in video monitors which use a CRT (cathode ray tube).

It wasn't until June of 1894, at age forty-nine, however, that Roentgen began to experiment with vacuum tubes. Roentgen began his own experiments with the Crookes Tubes and made several important observations. He noticed that he obtained better results when the tubes were highly evacuated, and he would often spend days evacuating a tube for his experiments. Also, he felt that if more current was applied to the tubes, cathode ray production would increase. Incidentally, he damaged many tubes during his experiments and thus began using heavier walled tubes.

While experimenting with the modified tubes in the fall of 1894, he noticed a faint glow in the room that precisely coincided with the discharge of the Crookes Tube. To be certain that the light was not coming from the tube or from the induction coil, he completely covered the Crookes Tube with cardboard so that no light could escape. He also covered the induction coil completely to eliminate any light caused by arcing within the machine. To his amazement, the glow still appeared in the darkened room.

The source of the glow turned out to be a screen made of a photographic material (barium platinocyanide) that was located several feet away from the tube. Some heretofore unknown invisible energy was being emitted from the vacuum tube and was traveling across the room, exciting the photographic screen. This was a highly significant observation since cathode rays had never been known to travel more than a few inches from the tube!

Roentgen experimented with the "new kind of ray" for nearly a year before he published his findings. On November 8 of 1895, at age fifty, he presented his paper "The First Communication," announcing his discovery.[4] He called the mysterious rays "x-rays," using the letter "x" (borrowing the mathematics symbol) to represent the "unknown" energy. Along with his paper he included several fascinating photographs that he had obtained by using the x-rays. These x-ray photographs included a metal compass, a box containing scientific weights, a double-barreled shotgun, and various types of metals.

Of all the published photographs, however, the one that immediately caught the attention of the media and general public was the x-ray

4. November 8 of the year this book was written (1995), marked the 100 year anniversary of this great discovery.

photograph of the hand of Bertha Roentgen, his wife. This famous photograph, showing all of the bones in her hand including the two rings on her finger, was to change the course of medicine forever.

In fact, Roentgen's x-rays proved to be the fastest spreading discovery of all time. Within months of the discovery, x-rays were commonly being used by physicians to diagnose bone fractures and to locate bullets in wounds from gunshots. X-rays were used so successfully that many overenthusiastic scientists predicted that the use of x-rays would eliminate the need for vivisection completely. In May of 1896, the first journals completely dedicated to x-ray photographs appeared in the United States. By January of 1897, many hospitals had x-ray departments.

In Germany, the first x-ray machines were being manufactured as early as February of 1896. These early machines required long exposure times (i.e., 4–20 minutes depending on the area being radiated) and, consequently, needed very much to be improved. It was then discovered that if a fluorescent screen was placed on both sides of the film, the exposure time would be reduced significantly. This finding led to today's widespread use of intensifying screens in film cassettes.

Because of the great demand for x-ray equipment, manufacturing companies all over the world began to produce x-ray machines. Roentgen never applied for a patent on his x-ray apparatus because he felt his discovery was for the public benefit and not for his own profit. As a result, a complete x-ray room could be purchased for less than $150.00. A portable x-ray machine, complete with a coil, condenser, two sets of tubes, a battery, and a handsome carrying case could be purchased for the price of about $15.00. X-ray tubes could be purchased for less than $4.00 a piece.

The public's fascination with x-rays and their effects was immense. Everyone wanted to see, for themselves, the effects that x-rays produced and to view their own bones. At the Electrical Exhibit in New York in May of 1896, Thomas Edison showed the world the first fluoroscope. With the fluoroscope, the internal structures of the body could be viewed directly on a fluorescent screen in real time, eliminating the usual wait involved for photographic plates to be developed. The person would simply place their hand (or body!) directly in the x-ray beam and the viewer would observe live motion. The immense popularity of the fluoroscope contributed to the widespread public use of x-rays.

Roentgen disapproved of this superficial use of his invention because he knew the effects of x-ray radiation had not yet been fully investigated. In fact, whenever he worked with x-rays, he would shield himself with a lead barrier from direct exposure to the rays as a precaution. Sure enough, soon after his discovery, news began to surface of injuries that were related to x-ray radiation. Among the reported injuries were burns, hair loss, and the swelling of radiated tissue. Most injuries occurred to doctors and demonstrators of the x-ray equipment.

Clarence Madison Dally, a glassblower and Edison's longtime assistant, regularly demonstrated the fluoroscope to the public. He later died from burns contracted from the continuous exposure to x-rays, becoming the first fatality as a result of radiation exposure. As of October of 1959, 359 deaths were attributed to excessive radiation exposure.

The early scientists who investigated x-rays were aware that tissue was affected, or changed, by radiation and warned the public of the potential hazards to tissue. This secondary characteristic of radiation, that it could destroy tissue, eventually led scientist to explore the field of ***Radiation Therapy.***

X-ray machines were commonly used during wartime. Their use significantly reduced the number of deaths due to battle injuries. In addition, many improvements in x-ray techniques were made as a result of the findings of the military. Surgeons noticed that the closer the x-ray tube was to the patient, the more the patient was likely to suffer radiation burns. Because of this finding, a minimum standard distance of 10 inches separation (between the x-ray tube and the patient) was established. Also, a maximum exposure time of 30 minutes was established for all patients.

Many significant changes in x-ray tube design were subsequently made. Roentgen designed the first focused x-ray tube in February of 1896. Thomas Edison found that better penetrating x-rays could be produced when cooling the x-ray tube by immersing it in oil. Also, higher voltages could be used while the tubes where immersed in oil. Many of the inventions of scientist Nikola Tesla, including his high voltage induction coils, helped to produce better quality x-rays.

By the early 1900s, x-ray tubes began to resemble present day tubes. In 1913, W. D. Coolidge of General Electric Company developed the hot cathode tube (thermionic tube) to allow for control of the amount of tube current (or mA). Coolidge also used tungsten in the construction of the target in order to provide greater tube stability. Ludwig Zehnder, who was Roentgen's assistant for many years, invented a metal-shielded, x-ray tube to protect doctors and technicians from stray radiation. The Phillips Company of the Netherlands eventually improved on this design and developed the "Metalix-Tube" which used a metal housing to surround the glass vacuum tube.

On December 10, 1901, Willhelm Conrad Roentgen received the Nobel Prize in Physics for his work with x-ray radiation. He spent the rest of his life exploring physical properties of matter and continued scientific experimentation with crystals even after his retirement in October of 1919, at age 75. He died on February 10, 1923.

The period from around 1910 to 1950 was considered the "golden age of radiology." During this time, x-ray tubes and all other x-ray devices advanced in design to their present-day status. Major advancements included: (1) the hot cathode tube (1913) for controlling tube current; (2) the first shockproof portable x-ray unit (1919) which employed the first shockproof high voltage cables; (3) the vibrating bucky (1923) which greatly reduced scatter radiation; (4) the high voltage rectifying switch (1910) which applied both alternations of the AC line to the x-ray tube, effectively doubling the radiation output; (5) oil immersed, rotating anode tube (1939) which greatly increased the output of the x-ray tube; (6) the tilting table (1943) which allowed radiologists to obtain that different anatomical views required for proper diagnosis; (7) intensifying screens (1912) which reduced the required exposure time significantly with the results of longer x-ray tube life and reduced patient dose; and (8) duplicating film (1914) which also greatly reduced exposure times. Most of the items listed above are used in today's radiographic rooms without significant change to the original design.

The general radiographic room has been around for 100 years and will continue to be the dominant mode of diagnostic imaging well into the future. Although advances in other medical imaging modalities (i.e., CT or MRI) have provided physicians with alternative methods to view the internal structures of the human body, general radiography studies still remain the primary diagnostic tool!

Chapter II

THE FUNDAMENTALS OF X-RAY PRODUCTION

In writing this book, the author assumes that the service engineer working in the field has some knowledge of x-ray theory. The objective here is to give a brief overview of x-ray production to ensure that there is a common starting point before discussing the servicing of x-ray equipment. If a more detailed discussion of x-ray theory is required, there are books available that deal specifically with the subject (see bibliography).

In a vacuum tube, if electrons are accelerated by some force, and then made to collide with a stationary object, energy will be released. The energy released is a result of the sudden deceleration of the electrons caused by the impact with the object. If the force applied to the tube is great enough to accelerate the electrons to a sufficiently high speed, and the object is made of the proper material, the energy released will mostly be in the form of heat energy, but with a very small amount of high frequency electromagnetic radiation. The frequency of this radiation is much higher than visible light (10^{18} hertz) and is called x-ray radiation (Figure 1).

Here is one of the most important concepts of x-ray production; a fact that should be permanently ingrained in the service engineer's memory. Specifically, during x-ray production ***99 percent o the energy released is given off as heat and 1 percent is in the form of useful radiation.*** This fact helps explain why x-ray tubes fail and will be discussed in detail later in the troubleshooting section of this book.

X-rays are produced in a thermionic vacuum tube called an ***x-ray tube.*** Thermionic x-ray tubes consist of a negative electrode (or cathode), and a positive electrode (or anode), enclosed in a glass envelope

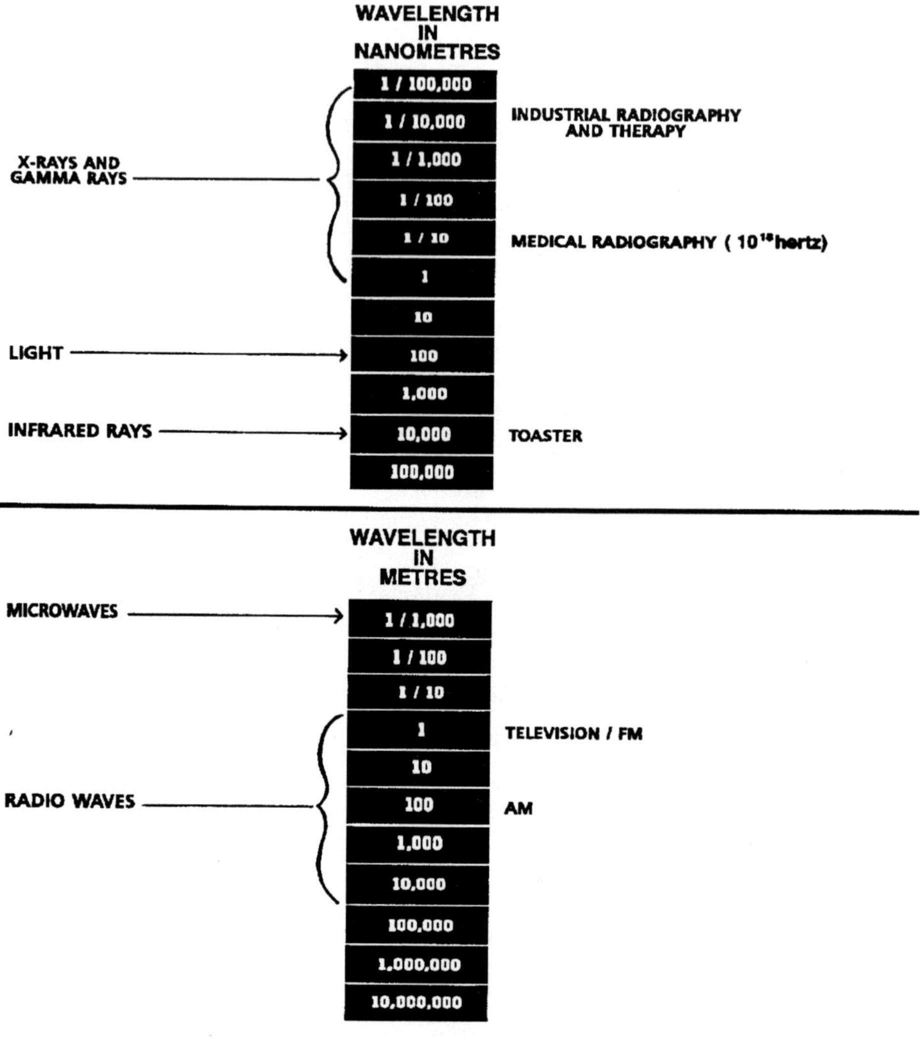

Figure 1. The electromagnetic spectrum.

(Figure 2). The glass envelope is surrounded by insulating oil and completely encased in a lead lined metal housing which contains a port or window.

In Figure 2, it can be seen that the cathode consists of a thin wire called a ***filament.*** This filament is similar to the filament in a light bulb but is surrounded by a ***focusing cup*** which concentrates, or focuses, the electron beam. The filament, when heated, will liberate free electrons

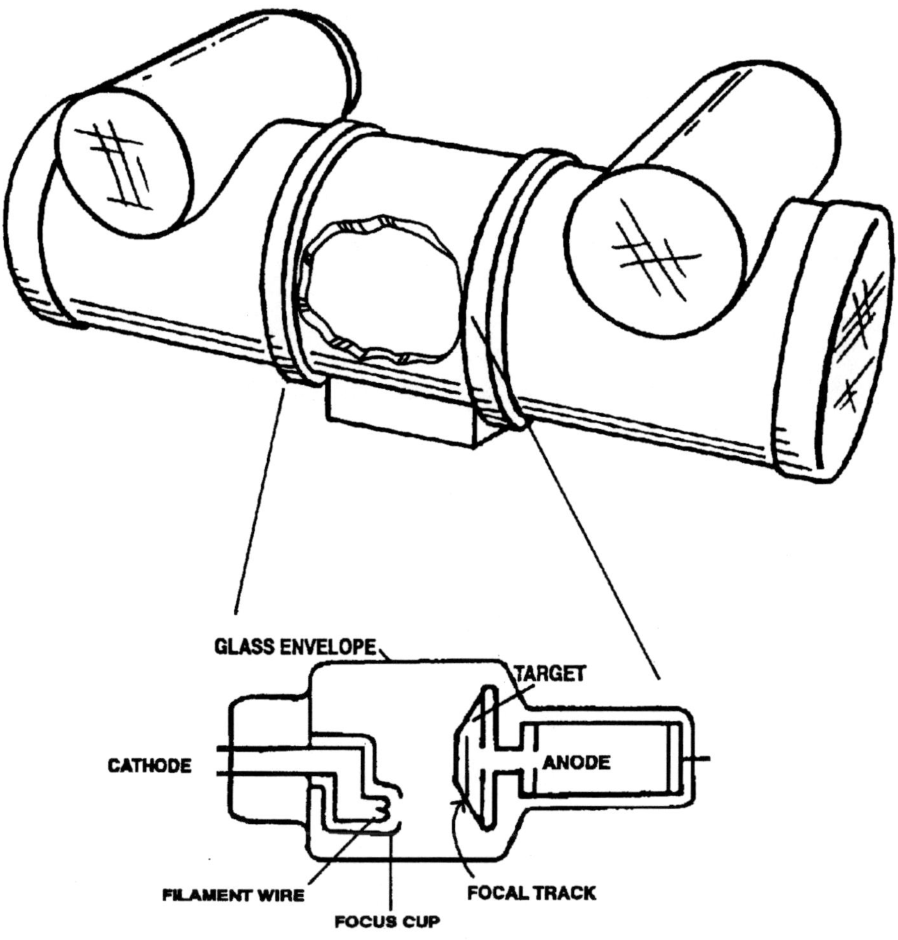

Figure 2. The x-ray tube. The x-ray tube glass envelope is completely surrounded by insulating oil within the tube housing.

that will eventually collide with the anode. Liberating electrons by the use of heat in the manner described above is referred to as ***thermionic emission.*** The amount of filament heating dictates the number of electrons that will be liberated, and consequently, the amount of x-ray tube current (or mA). Controlling the heating of the filament is referred to as ***filament control.***

The anode or positive electrode contains the target, which is made of a special material (e.g., tungsten) that has a high melting point and a high atomic number. High atomic numbers are used to ensure that the x-ray beam will contain photons of a high energy level. When the electrons collide with the target, x-rays are emitted in all directions from the target. To ensure that radiation will only exit through the port of the tube, the metal housing of the x-ray tube is completely lined with lead.

The force that causes the electrons to travel at a high rate of speed is a ***high voltage potential*** that is applied across the anode and cathode of the x-ray tube. The amount of high voltage not only affects the quantity of radiation but also affects the quality or penetrating ability of the x-rays (see Chapter VIII). Regulating the high voltage is called ***kVp control.***

In short, in order to produce x-rays in a thermionic tube there must be: (1) a filament voltage supply applied to the filament wires of the tube; and (2) a high voltage potential applied across the tube. If either of these supplies is missing, x-rays cannot be produced.

Chapter III

THE X-RAY SYSTEM

A typical x-ray system consists of an ***x-ray control,*** a ***high voltage transformer unit,*** and an ***x-ray tube*** (Figure 3). The high voltage transformer unit and x-ray control are often considered as one unit by many service engineers and together are referred to as the ***x-ray generator.*** The x-ray tube and high voltage transformer unit are connected by special ***high voltage cables.*** Some x-ray generators, although rare, combine the high voltage transformer and x-ray tube together as a tube head assembly. In this case there are no high voltage cables required.

THE X-RAY CONTROL

Three factors control the quality and quantity of x-rays produced. They are: ***kVp, mA,*** and ***time.*** These three factors are regulated in the X-ray Control or Generator. KVp refers to the amount of high voltage, or peak kilovoltage, applied to the x-ray tube. MA is the amount of current, or milliamps, that is actually passing through the tube. Time refers to the period during which the x-ray tube is energized and emitting x-ray radiation (i.e., actual exposure time). Each of these three factors has an important effect on image quality which will be discussed in detail later in this book (Chapter VIII). For now, our discussion will concentrate on the common types of x-ray generators.

There are four basic types of generators: single-phase, three-phase, constant potential, and high frequency. Each generator type generates and regulates (or *controls*) the kVp by different means. Consequently, each generator type produces a distinct kVp waveform and can be eas-

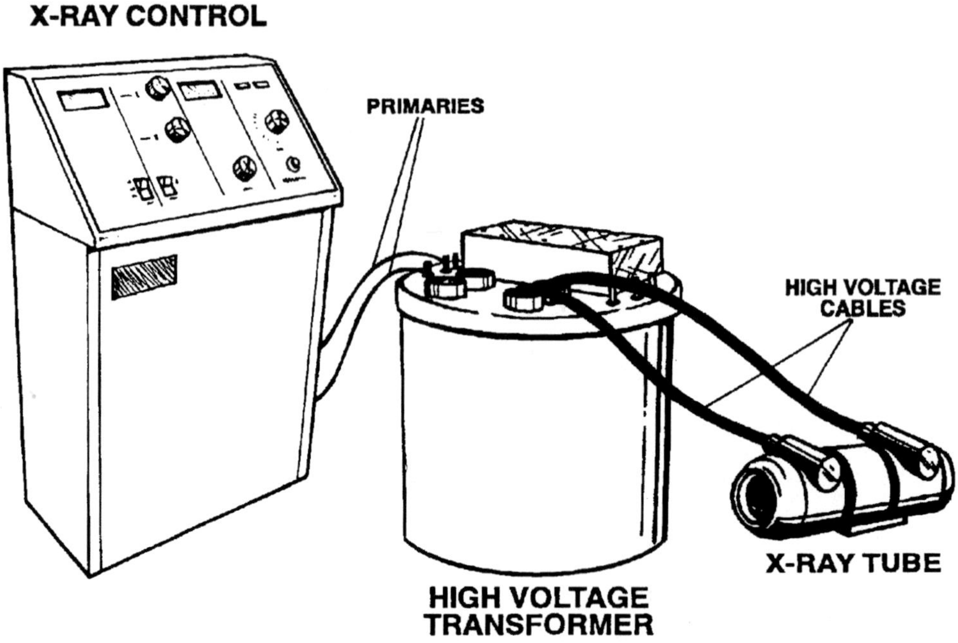

Figure 3. A typical x-ray system.

ily identified simply by viewing that waveform invasively on an oscilloscope.

KVp Control

The Single Phase Generator

The ***single-phase generator*** was the first type of x-ray generator manufactured and is the most basic in design. Although they are still commonly used today, they remain the least efficient generator in producing radiation and expose the patient to significantly more radiation per exposure than the other types of generators. However, because they require only a single-phase power line, are physically smaller in size, and are the most affordable of all generators, many customers, especially outpatient facilities, consider this type of generator the generator of choice. Furthermore, because of the basic circuit design of single-phase generators, there are fewer components

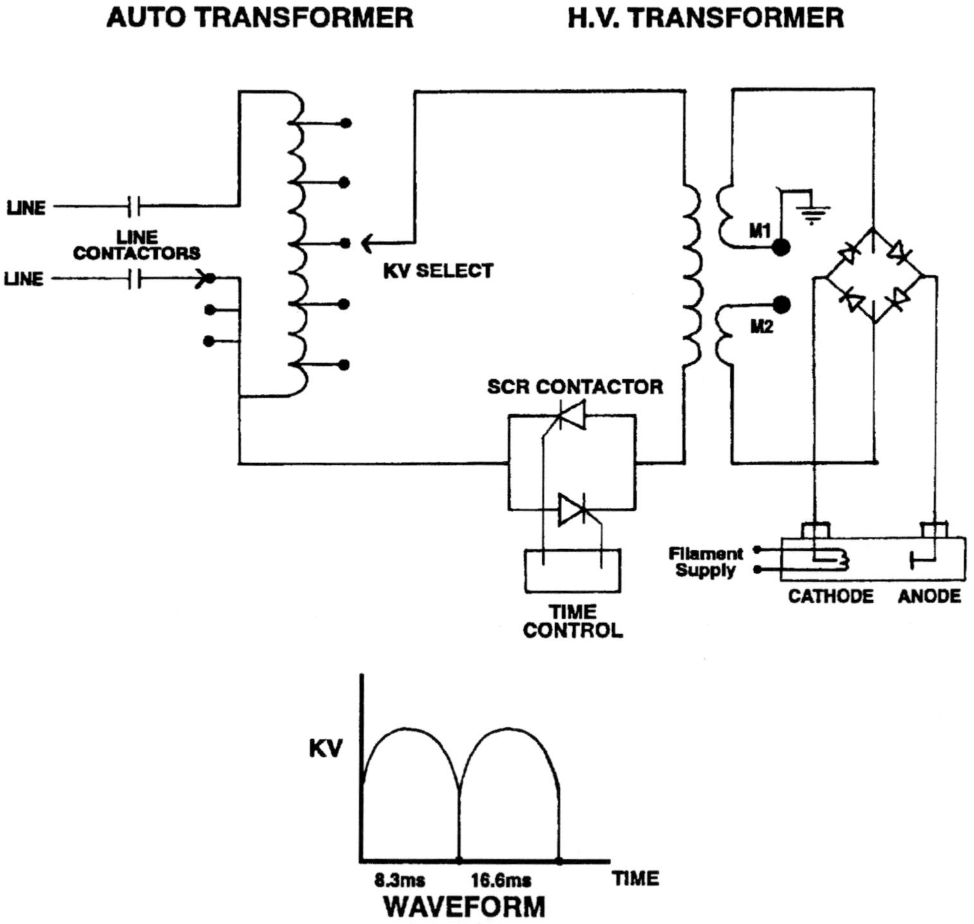

Figure 4. Single-phase generator and output waveform. Single-phase generators produce an output waveform with 100 percent ripple.

involved in the production of radiation, and, therefore, servicing these units is fairly straightforward.

In this design, a primary voltage corresponding to the selected kVp is outputted from the autotransformer and applied to the primary coil winding of the high voltage transformer (Figure 4). The primary voltage is then stepped up through the transformer to the desired secondary high voltage, is rectified (i.e., converted to DC), and then applied to the x-ray tube.

From the diagram we see that a single phase generator is easily identified by its characteristic, 100 percent ripple, DC waveform. Since useful radiation is produced only during the peak portion of the waveform (approximately 30% of each pulse), it is clear that this type of generator is not very efficient at x-ray production.

The Three-Phase Generator

Three-phase generators are a great improvement over single-phase generators in that they can deliver more power, are more efficient at producing x-ray radiation, and provide better time control. Three-phase generators require three different line phases for input power, each phase being 120° out of phase with the other two. Instead of a single autotransformer, these generators have three separate, usually motor driven, autotransformers–one for each phase. The primary voltage that is derived from the autotransformers is applied to the primary of a special three-phase, high voltage transformer.

There are two types of three-phase, high voltage transformers used in these x-ray generators. The type of transformer, and its corresponding output waveform, depends on the type of secondary winding configuration employed (Figure 5). The secondary could be configured with two separate ***Delta*** winding, two separate ***Wye*** (or ***Star***) windings, or the secondary could have a ***Delta-Wye*** configuration. If the same winding configuration is used in the secondary, then it is termed a ***six-pulse generator.*** If a ***Delta-Wye*** winding configuration is used in the secondary, then this designates a ***twelve-pulse generator.***

A six-pulse generator produces a DC output waveform that has 13.5 percent ripple. This is a significant improvement over single-phase generators. A twelve-pulse generator, however, produces an output waveform with only 3.4 percent ripple. By examining the kVp waveforms, it becomes evident that a twelve-pulse system is more desirable because of its near "constant potential" DC waveform.

Three-phase generators provide many advantages over single-phase units. The increased output efficiency of these generators allows for much shorter exposure times, as compared to a single-phase generator using the same kVp setting. In addition, much higher exposure techniques can be used at short exposure time settings with three-phase systems. Also, the exposure time is more accurately controlled in three-phase generators (see time control), a requirement for serial radiography and phototiming techniques. For these reasons, three-phase generators are frequently used for angiographic and spot filming applications.

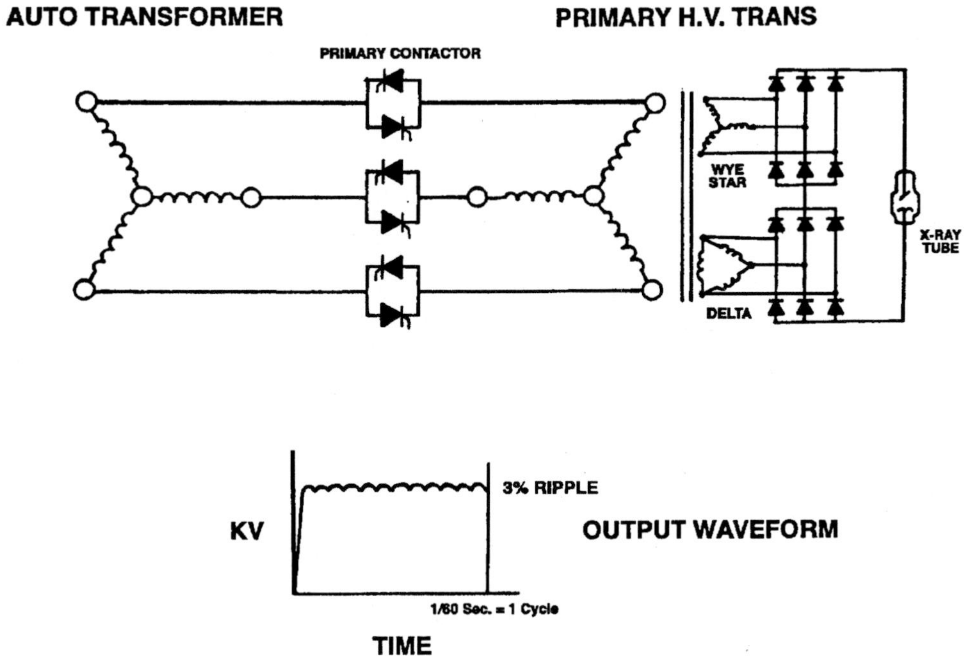

Figure 5. Three-phase generator (12 pulse) and output waveform. Three-phase (12 pulse) generators produce an output waveform with only 3.4 percent ripple.

The Constant Potential Generator

A special version of a three-phase generator is called a ***constant potential generator.*** Here, additional switching circuitry is employed in the high voltage secondary circuit to achieve the accurate exposure time control required for high-speed filming applications. Because this type of generator utilizes a high voltage "switch," the exposure times are precisely controlled and, in addition, the resulting kVp waveform is a DC square wave that is essentially ripple free (Figure 6). With a true square wave output, the kVp remains constant for the duration of the exposure, hence the name "Constant Potential." Constant potential generators are the most expensive generators available and are used in Cath Lab and Special Procedure applications.

One way the fast switching of high voltage is accomplished is by the use of ***tetrodes*** in the high voltage secondary. The tetrodes, located in a separate tank called the ***tetrode tank,*** are wired in series with the x-ray tube. An external exposure signal, sent from an imaging device such as

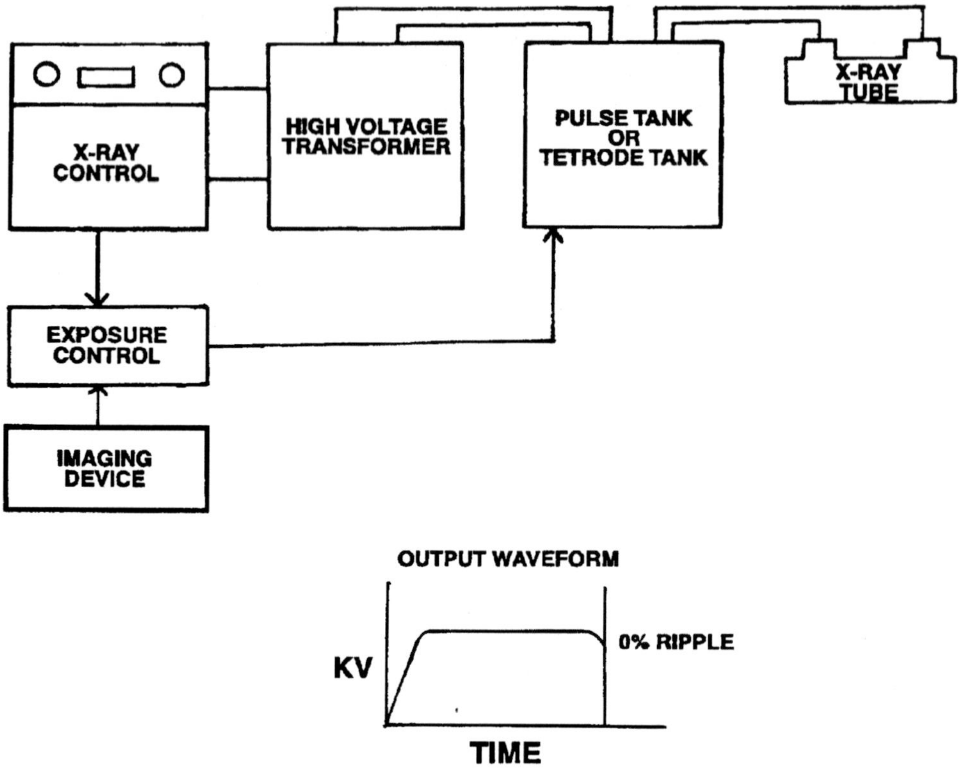

Figure 6. Constant potential generator block diagram and output waveform. Constant potential generators produce an output with no ripple.

a cine camera or film changer (or handswitch in radiographic mode), turns the tetrodes "on" and "off" which, in turn, allows the x-ray tube to conduct.

Another way to accomplish fast switching is by the use of a ***grid-controlled x-ray tube.*** With a standard x-ray tube, the grid of the tube (i.e., the focusing cup) is normally shorted to the common reference point. With a grid-controlled x-ray tube, the grid is held at a negative potential (in relation to the cathode) and is essentially switched "off" even when high voltage is applied to the x-ray tube. During an exposure, the grid voltage is pulsed "on" and "off" and, in turn, the tube is turned "on" and "off" by pulses sent from a ***pulse tank.*** As with the tetrodes, the pulse tank receives an exposure signal from an imaging device to initiate an exposure.

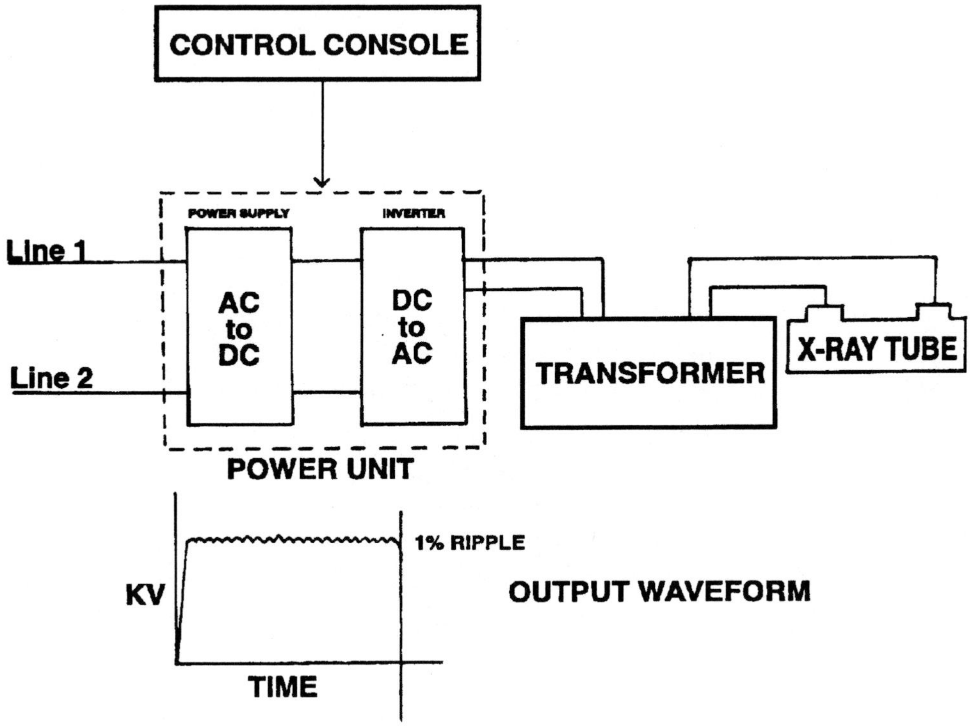

Figure 7. High frequency generator block diagram and output waveform. High frequency generators produce an output waveform with less than 1 percent ripple.

The High Frequency Generator

High frequency generators are the latest type of x-ray generator and are, by far, the most efficient at producing x-ray radiation. This type of generator can produce a high voltage square wave that is essentially ripple free, providing uniform radiation for the entire exposure. Another advantage of this type of generator is that it can operate with either a single-phase or a three-phase input line. By looking at the simplified block diagram (Figure 7), it is clear that high frequency generators utilize a unique method for producing high voltage.

The most significant difference with this type of generator is that the ***incoming line voltage is immediately converted to direct current,*** providing a constant source to the power inverter section. This design provides the advantage of improved kVp regulation. With all other generator designs, the kVp is derived directly from the AC line. Consequently, any fluctuations in the incoming power line will cause a corresponding fluctuation in the kVp waveform. With high frequency generators, the high quality DC supply maintains a stable, filtered DC output regardless of any inconsistencies with the AC line.

The regulated output from the DC supply is fed directly to the power inverter section of the generator. This section consists of an "inverter" which converts the DC supply voltage into high frequency pulses, thereby simulating an AC waveform. The inverter sends these pulses to the high voltage transformer which then steps up the voltage to the correct level of high voltage that will be delivered to the x-ray tube.

Another advantage of using a high frequency supply is that significantly smaller components can be used in the high power circuitry.[5] This means that the high voltage transformer, which usually takes up a large area, can be placed inside the control cabinet along with the other components of the generator control. This greatly reduces the amount of space needed for an installation. In many cases, the size limitation of the room is a major consideration in the purchase of radiographic equipment. In such situations, then, high frequency generators are the ideal choice.

Other advantages of high frequency generators are that they are much easier to calibrate, they will hold a stable calibration for a longer period of time, and provide more accurate time control as compared to the other generators. This is mainly due to the fact that all high frequency generators utilize feedback circuitry to monitor (in real time) the actual kVp, mA, and time of the exposure. Because of the sophisticated signal monitoring, they are also easier to troubleshoot and repair. Most high frequency generators have error detection circuitry that can isolate a failure and even provide a message on the console display to alert the operator of a problem. High frequency generators have improved greatly in design and are used for all applications including angiography.

---

5. Basic electronic theory demonstrates why this is so. Referring to the formulas for capacitive and inductive reactance $X_C = 1/2\pi FC$ and $X_L = 2\pi FL$, it can be seen that if all other factors are held constant, increasing the frequency would require a corresponding decrease in capacitance and inductance respectively and, therefore, smaller components.

MA Control

In addition to controlling the kVp, the x-ray generator also controls the amount of tube current flowing through the x-ray tube. MA is the term used to refer to the actual current passing through the x-ray tube and is directly controlled by the amount of current going through the filaments of the x-ray tube. For a given increase in filament current there is a proportional increase in mA (kVp held constant).

A typical mA control circuit consists of a filament voltage regulator (or stabilizer) that provides a constant supply voltage to the filament circuits as the load varies, and a means of controlling the voltage that is applied to the filament transformer (Figure 8). Filament voltage regulation is accomplished in different ways ranging from simple resistor tap selection to frequency control.

The difficulty with mA control involves a phenomenon in x-ray tubes called ***space charge effect.*** Consequently, more circuitry is required to provide accurate mA regulation. The process of correcting for the space charge effect is called ***space charge compensation.***

Space charge is defined as the cloud of electrons that surrounds the filament when it is being heated. The size of the cloud is proportional to the degree of filament heating. The problem when dealing with the space charge phenomena is when we consider the kVp that is applied to the tube. If the filament current is held constant, an increase in kVp will cause a proportional increase in tube current (mA) because of the greater "pull" exerted on the electrons in the cloud. Conversely, a decrease in kVp will cause less "pull" on the cloud of electrons and this will reduce the actual tube current. Space charge compensation is utilized specifically to maintain a constant mA through the full range of kVp settings.

The space charge compensation circuit is connected to the filament circuit at a point after the mA adjusting resistors. This circuit monitors the selected kVp and then applies the appropriate compensating voltage that is needed to maintain the selected mA at a constant level. At higher kVp settings, the space charge compensation circuit will reduce the voltage that is applied to the filament transformer, and at lower settings the compensation voltage will be increased. Generally, there is no space charge compensation applied at or near 80 kVp. Because of this fact, the preliminary mA calibration is performed at this kVp station (see x-ray calibration in Chapter VI).

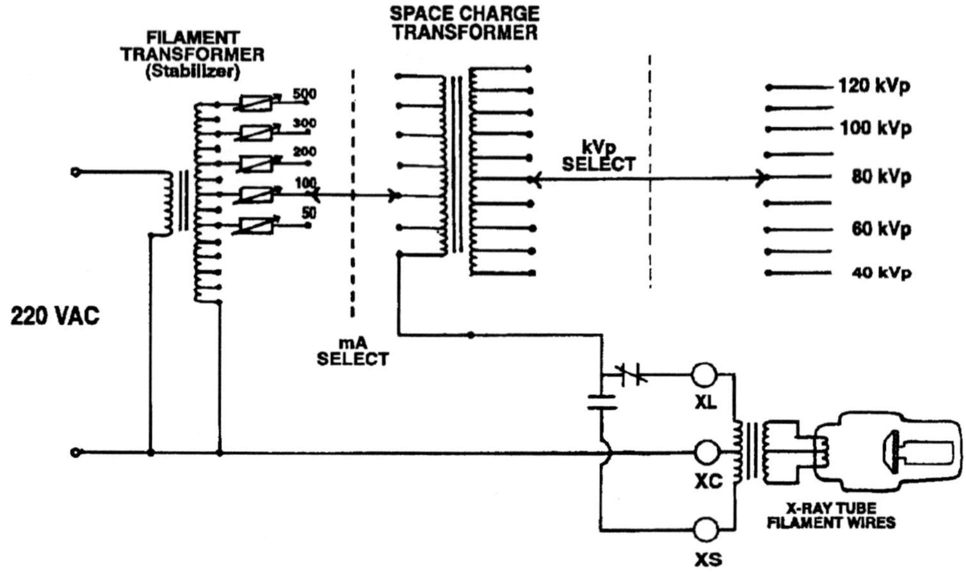

Figure 8. Filament control circuit. A correction voltage is applied to the filament wires to compensate for the space charge effect. The correction voltage is determined by the kVp selected.

The Falling Load Generator

A primary cause of x-ray tube failure is related to filament damage caused by excessive tube currents. The constant use of high mA's will definitely shorten the life of an x-ray tube (see Chapter IX). To address this problem, some manufacturers of x-ray equipment developed a special type of generator for the specific purpose of increasing tube life. This type of generator is called a ***falling load generator*** and is named after the actual mA waveform at it produces.

Whereas, all other x-ray generators are distinguished by their kVp waveform, falling load generators are identified by the mA waveform. With falling load generators, the exposure begins with the highest mA allowable for that particular x-ray tube and technique setting. After a set period of time (e.g., 10 ms) the mA is significantly reduced producing the distinctive falling load mA waveform (Figure 9). The mA waveform continually "falls off" or decreases during the entire exposure, thus applying much less stress to the x-ray tube filaments.

This falling load design will increase x-ray tube life, but at a cost: longer exposure times. Much longer exposure times are required to

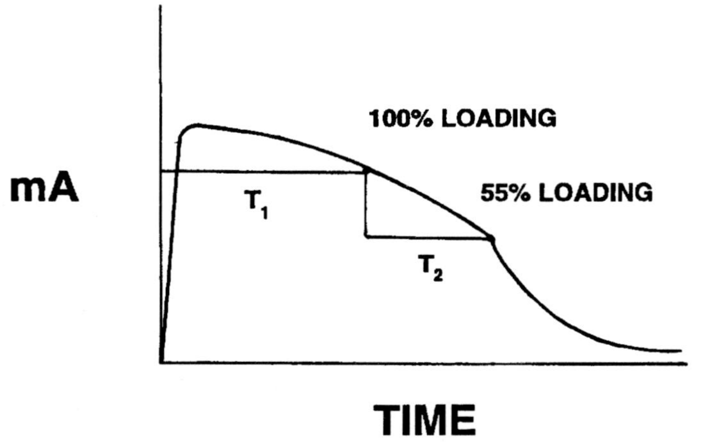

Figure 9. Falling load generator mA waveform. After a programmed time interval (t_1), the mA is reduced to a lower value for the remainder of the exposure. This design effectively extends x-ray tube life.

compensate for the greatly reduced tube current. The long exposure times cause a reduction in image quality mainly because of patient motion (see Chapter VIII). This major disadvantage partly contributed to the decline in use of the falling load generator. Moreover, because of advances in x-ray tube design, film and intensifying screens, and the increased popularity of high frequency generators, falling load generators are no longer in demand.

Time Control

Time Control is the third major function of the x-ray generator and is accomplished in different ways depending on the type of generator. The timer circuitry must initiate the x-ray exposure and then accurately stop it at the selected time. Time control is accomplished by alternately "switching" the primary voltage of the high voltage transformer "on and off."

In the early x-ray units, the actual switching was done by large relays called ***contactors.*** Today, SCR's (silicon-controlled rectifiers) are widely used instead, although the term "contactor" is still used. SCR's are used because they provide transient free switching. These contactors are signaled "on and off" by the SCR control, which receives its signals from the timer circuitry.

Single-phase units have historically used analog timers, although most newer units are now utilizing digital timing. With single-phase, timing circuits, the AC waveform of the primary circuit is monitored for the zero crossover points. The zero cross-over points are converted into timing pulses and are used to initiate, measure, and stop the exposure.

The diagram in Figure 4, on page 15 demonstrates that each alternation of the AC sine wave (or 1/2 cycle) is 8.33 milliseconds in duration, and, therefore, the shortest possible exposure obtainable with a single-phase generator is precisely 8.3 milliseconds (1/120 s). This exposure time, however, is not nearly fast enough for phototiming and spotfilming applications which often require exposure times that are less than 5 milliseconds. Even for conventional radiography, exposure accuracy is reduced when using short exposure times with single phase generators.

Three-phase generators also use either analog or digital timers depending on the vintage of machine. However, the important issue with respect to three-phase timers is the method used to terminate the exposure. Two methods are commonly used: ***self-commutation*** and ***forced commutation.***

X-ray generators that employ self-commutation have a contactor (back to back SCR's) for each line phase. With self-commutation, the exposure will terminate at the next zero cross-over point following the exposure stop signal (Figure 10). The diagram shows that two line phases must "turn on" to make a return path for the primary circuit. As a result, it takes 2.7 milliseconds for the anode plus 2.7 milliseconds for the cathode to turn "on." This means that the shortest possible exposure time is 5.44 milliseconds.

Although an improvement over single-phase units, the response time is still not quick enough for today's high-speed, film/screen combinations. Many x-ray units that employ self-commutation timing control can be outfitted with an add-on external unit that will "force" the SCR's "off" when an x-ray stop command is generated, allowing for exposure times as short as one millisecond. These units are referred to as ***forced extinction packages.***

The second type of three-phase timer utilizes ***forced commutation*** to terminate the exposure. The forced commutation design solves the problem of exposure termination by utilizing two SCR contactors and a bridge network at the midpoint of the primary winding of the high

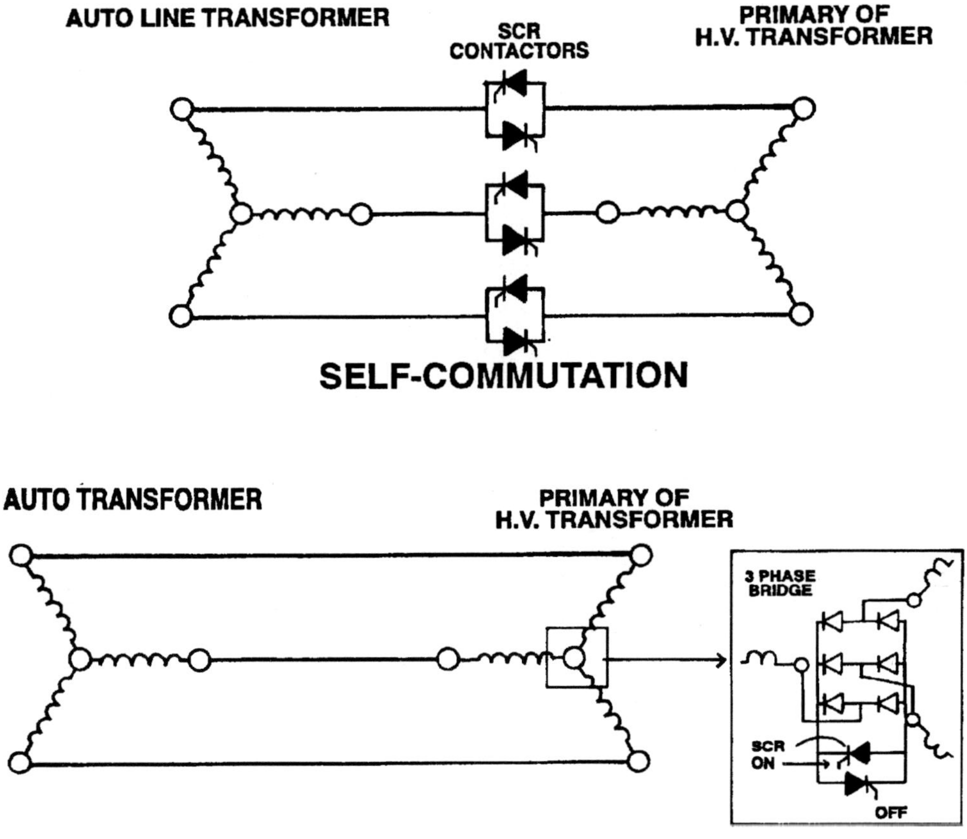

Figure 10. Self-commutation vs. forced commutation. (Top): Notice that two line phases must be switched on for current to flow. (Bottom): One contactor, located at the midpoint of the primary winding of the high voltage transformer, controls current flow.

voltage transformer (Figure 10). One contactor completes the path of the primary circuit which then initiates the exposure. The second contactor forces the SCR that initiated the exposure; "off" by reversing the anode-cathode polarity. Exposure times as short as 1 millisecond are possible with forced commutation.

Core Memory Circuits

Another important function of the time control circuits is to provide ***core memory detection.*** The iron core of high voltage transformers (or any solid core transformer for that matter) must see alternate cycles of polarity in order to function properly. If, for example, the positive cycle of the AC waveform was the last cycle to pass through the primary of the transformer, the next exposure must begin with a negative alternation. If the same polarity were to follow, the iron core would become "saturated," a condition that severely affects the performance of the transformer. Core saturation will result in reduced efficiency of the transforms and will usually cause a serious high voltage failure (see Chapter IX).

Core memory circuits prevent the iron core from becoming saturated by monitoring the polarity of the line and, consequently, starting the next exposure with the correct (opposite) alternation.

Other X-ray Control Functions

Besides providing kVp, mA, and time control, the generator also provides several other important functions. In order to ensure safe operation of the x-ray tube, all generators are equipped with ***tube protection circuitry.*** In the most basic generator designs, tube protector circuits will prevent the operator from selecting a technique that exceeds the safe operating limits of the x-ray tube. When a technique is selected that exceeds the tube limitations, an "exposure hold" condition is created accompanied by a visual "technique over" indication. The technician must reduce the technique settings on the ray control before an exposure will be allowed.

More sophisticated microprocessor-controlled units can display an error message if a technique exceeds the tube ratings. In addition, microprocessor-controlled units will usually have other special features such as a ***heat unit calculator*** that displays the amount of heat units available for use, or ***tube loading*** indicator that displays the percentage of the tube's maximum output that is being used for a given technique.

At installation, the service engineer must program the generator with the exact tube specifications in order for this circuitry to operate properly. Some generators provide preinstalled software that contains data on the specifications of the most commonly used x-ray tubes. This software is preprogrammed at the factory and, during the installation, the engineer will simply enter into a service mode from the console and select the appropriate tube.

Rotor control is another important function of the x-ray generator. Because low-speed starters are relatively smaller in size, they are always located within the x-ray generator control cabinet. On the other hand, high-speed starters are physically larger and are mounted in a separate cabinet.

Both high and low speed rotor controllers must provide: (1) the proper starting voltage to initiate rotation; (2) a means to adjust the "boost" time that the starting voltage will be applied to the stator windings; (3) a lower running voltages to maintain rotation; (4) the proper phase shift capacitor for the tube that is being used; and (5) an exposure interlock circuit to ensure that the tube is rotating before an exposure is attempted.

An ***AEC control,*** optionally available with most x-ray generators, is also located in the generator control. If an AEC option is not available for the particular model generator, a compatible external AEC control unit can be purchased from any one of many different manufacturers and interfaced to most x-ray units. When properly calibrated, the AEC will control the exposure time by sensing the radiation that passes through the patient. Automatic exposure control is a much more efficient way to obtain consistent x-ray films. Consequently, many radiologists request this option.

All generators also provide ***mA metering*** and ***safety circuits.*** These circuits are usually connected at the midpoint of the secondary of the high voltage transformer. The midpoint, usually referred to as M1, M2, is referenced to ground and provides a convenient point to safely monitor the actual x-ray current flowing through the tube. The analog mA meter on the front panel of many generators is connected in series to this point, and this is also the point where an external mA/mAs meter is installed during calibrations.

THE HIGH VOLTAGE TRANSFORMER UNIT

As stated earlier in this chapter, the filament wires of the x-ray tube must be heated to liberate the electrons used for x-ray production. A low voltage, regulated power source, in the range of 6–12 VAC, applied to the filaments in the cathode of the x-ray tube, provides this heating. Also, a high voltage source in the range of 40 kilovolts to 150 kilovolts is needed to accelerate the electrons with sufficient speed so that when they collide with the target, x-rays will be produced. The ***filament***

transformer (for the low voltage source) and the ***high voltage transformer*** (for the high voltage source) are both located in the high voltage transformer unit. Both transformers are completely immersed in insulating oil.[6]

Other components located in the high voltage transformer unit include the *stick rectifiers* (or valve tubes in early vintage units) and the *high voltage receptacles.* Some transformer units provide additional sets of receptacles for energizing two or more x-ray tubes. These wells are activated by a solenoid or a motor-driven switch which is also located in the transformer unit.

If kVp and mA feedback signals are provided with the x-ray generator, the feedback detection circuitry is located within the transformer unit. Also, there are convenient test points to monitor the primary filament current and x-ray tube current (mA) at the transformer unit.

THE X-RAY TUBE

As mentioned, the actual production of x-ray radiation occurs inside the x-ray tube. There are many different types of x-ray tubes available today from a variety of manufacturers. These tubes vary in anode construction, focal spot size, maximum heat capacities, and tube housing style. Each type of x-ray tube must be correctly matched to the specific radiographic equipment being used and for the intended clinical use. The service engineer should, therefore, be familiar with the characteristics of the various types of x-ray tubes so that he or she can correctly advise the radiology administrator or the equipment dealer as to which tube would be the best choice for the intended application.

X-ray tubes are designed with either a *stationary mode* or a *rotating anode.* A stationary anode is constructed of a copper stem onto which the target is mounted. Copper is used because of its ability to dissipate heat as it accumulates on the target during an exposure. Because the anode remains stationary at all times, the stream of electrons emanating from the filament will impact the target at only one area (or focus) of the target. This less than ideal situation will result in a limited tube

---

6. Since the secondary of the filament transformer is directly connected to the cathode of the x-ray tube, it is at a high potential and therefore must be insulated.

capacity, shortened anode life, and a noticeable increase in effective focal spot size during the life of the tube.

The main advantage of the stationary anode is that it is much cheaper to manufacture because there is no need for the additional circuitry required to provide anode rotation. Stationary anodes, although commonly used for radiography years ago, are now only used in lower output x-ray units and a few smaller portable units.

Rotating anodes are preferred because of their greater heat capacity, higher radiation output, and longer target life. As the anode rotates at a constant speed, the electron beam now impacts an area along the entire surface of the target called the ***focal track.*** Since the heat is dissipated over a much greater surface area, much higher technique factors can be used for short exposure times. In addition, the target will last much longer because it is heated more evenly than with stationary targets.

Rotating anode tubes are, however, more expensive to manufacture because of the special construction and additional components required to produce anode rotation. The rotating anode basically is a single phase induction motor with a target attached. This induction motor consists of a rotating part called the ***rotor*** and a number of stationary electromagnets (or field windings) that are together referred to as the ***stator.*** The target is directly attached to the rotor of the motor.

The stator in all x-ray tubes utilize what is termed a ***split phase configuration*** to achieve rotation (Figure 11). This type of stator requires a capacitor to be placed in series with one of the three stator windings (the "start" or phase-shift winding). The capacitor causes a phase shift to occur between the main and phase winding leading from the single phase power supply, effectively creating the two-phase supply required to initiate rotation.

The induction motors used in x-ray tubes vary in design among manufacturers, and can be distinguished by the number of field windings used. The service engineer can measure the impedance of the stator windings to determine what type of stator is being used. The three most common types of stators used are the R (or standard), Q (or low impedance), and E (or balanced) stators (see Appendix E).

The actual rotation of the anode is controlled by the ***rotor controller*** or ***starter,*** located in the x-ray control (see above). The starter is simply a single phase AC motor control unit that initiates and maintains rotation of the anode. There are two types of starters: ***normal or low speed***

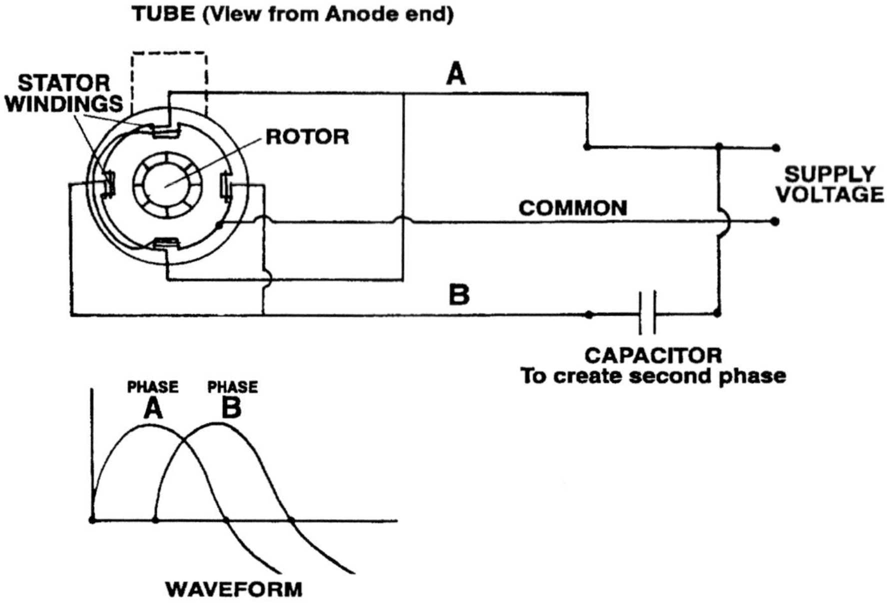

Figure 11. Split-phase stator configuration. The capacitor used to create the two-phase supply is physically located in the rotor controller.

starters and ***high speed starters.*** A low speed starter will induce an anode rotational speed of 3,300 rpm's and is mostly used for general radiography applications. High speed starters can induce a rotation of 10,000 rpm's and are used in applications where high x-ray exposure techniques are needed (i.e., special procedures and cine applications). Using an x-ray tube in high speed mode will allow for a greater accumulation of anode heat which, in turn, increases the output ratings of the tube.

The starter controls the speed of anode rotation by first applying a ***starting voltage*** to the stator windings for a calibrated period of time, called ***boost time,*** and then switching to a lower ***running voltage*** for the remainder of the exposure. The starting or boosting voltage will quickly get the rotor up to the correct speed and the running voltage simply maintains that optimum speed. The values for the start and run voltages are adjusted during the installation and the anode rotational speed is then verified. The boost time is adjusted on all rotor controllers to account for anode weight and bearing condition.

In addition to initiating and maintaining anode rotation, the rotor controller also provides another important function. All rotor controllers have an ***exposure interlock circuit.*** This special circuit detects the rotation of the anode and then sends an exposure release signal to the generator to allow the exposure to occur. Most exposure interlock circuits utilize current sensing circuits that operate in a similar manner. As the anode rotates, the current passing through one of the stator leads is sensed by the winding of a special (current sensing) transformer. This transformer then outputs a ("rotor ready") confirmation signal to the exposure control logic of the generator to allow the exposure to proceed.

The part of the anode where actual x-ray production occurs is called the target. The target is usually constructed of tungsten, rhenium, graphite, molybdenum, or a combination of these materials. These metals are used because of their high melting points and high atomic numbers. The higher the atomic number, the more efficient the target becomes in producing x-rays.

The diameter of the target is an important consideration in x-ray tube design. A larger diameter target of four inches, for example, will have a greater (as much as 30% higher) heat capacity and higher short time exposure rating than a three inch diameter target. The improved heat ratings can be attributed to the larger ***focal track*** which helps dissipate the heat. Anode heat ratings are specified in ***heat units,*** or ***HU*** for short. For single phase units, one heat unit is equal to ***1 kVp X 1 mA X 1 second.*** The heat unit was chosen as the unit of measurement because it is much easier to work with than the conventional ***watt-second.*** Typical anode heat ratings for general radiographic x-ray tubes range from 300,000 to 400,000 HU.

The ***target angle*** will directly determine the x-ray field coverage, the heat rating of the tube, and the effective focal spot size of the x-ray tube. A smaller target angle will produce a smaller effective focal spot size but with reduced film coverage (see Appendix B). Also, a smaller target angle will allow greater short time exposure ratings of the tube.

A larger target angle, on the other hand, will increase the x-ray field coverage, produce a larger effective focal spot size, and will reduce the exposure rating of the tube. The goal when selecting an x-ray tube is to choose a minimum target angle that will achieve the film coverage desired.

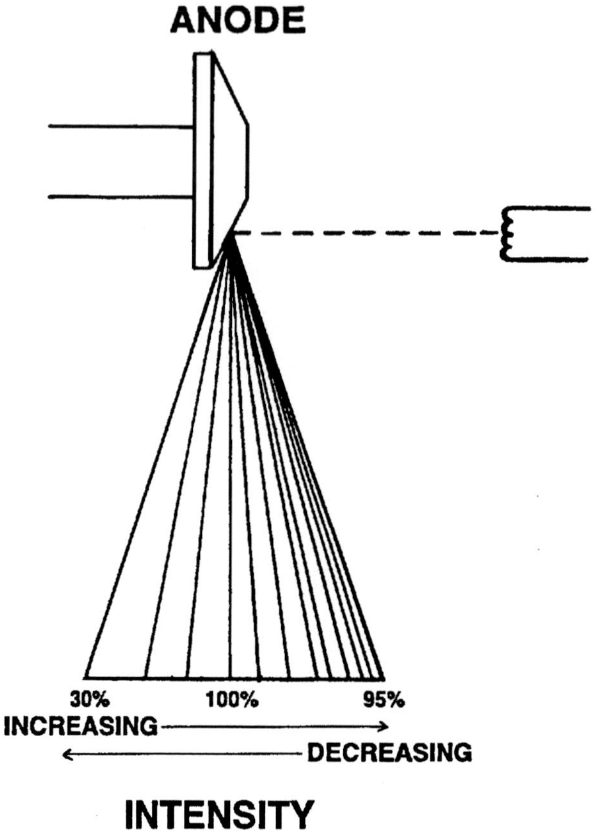

Figure 12. The heel effect. The x-ray beam is attenuated as it passes through the thickness of the target material.

Heel effect is a phenomenon of x-ray tubes that must also be taken into consideration when selecting a target angle. Heel effect is defined as *the variation in exposure rate with the angle of emission of the radiation from the focus.* The diagram below (Figure 12) demonstrates that the radiation gradually decreases in energy in the direction towards the anode. This is due to the greater attenuation of the x-rays as they pass through the thickness of the anodes "heel." A smaller target angle produces more heel effect.

The engineer will compensate for the heel effect during the installation by mounting the x-ray tube in an orientation (in respect to the image receptor) that minimizes the heel effect. As a rule, the anode is usually located at the head end of the table or receptor.

The focal spot size of the x-ray tube must also be taken into account when choosing an x-ray tube. The focus of the tube, or ***effective focal spot size,*** is directly related to the size of the filament wire and, as mentioned earlier, the target angle. Most of the x-ray tubes manufactured today are dual focus tubes which provide both a small and large focal spot.

For general radiography, the small focus of the x-ray tube is usually in the range from 0.6 mm to 1.0 mm, whereas the large focal spot size ranges from 1.0 mm to 2.0 mm. For applications requiring fine detailed imaging (i.e., vascular imaging), smaller focal spot sizes are required. The smallest focal spot sizes used in radiography are used in mammography, with 0.1 mm/0.3 mm focus being the most common focal spot combination. The drawbacks of using a smaller focal spot size, however, are that the overall heat ratings of the tube and short time exposure ratings are significantly reduced.

The type of ***x-ray tube housing*** is another important consideration when choosing a tube. The heating and cooling ratings of the tube housing are directly determined by the type and size of housing. These ratings, usually given in graphic form, show the maximum heat accumulation allowed per unit of time for the particular housing. From the graphs (included with the tube rating information) the engineer can determine how quickly the housing can dissipate heat. Typical heat storage capacities for x-ray tube housings used in general radiography are rated well over a million heat units!

The heating/cooling ratings are not only dependent on the size of the housing, but also by the method of cooling that is employed. X-ray tubes can be ***air cooled*** (with or without an external fan), ***oil cooled,*** and even ***water cooled.*** Air cooled tubes are most commonly used for general radiographic applications. Many air cooled tubes are equipped with an air circulator (i.e., box fan) which greatly increases the heat rating of the tube housing. Oil and water cooled tubes have even greater heat capacities and, consequently, are used for special high technique applications. The oil (or water) in the housing is circulated by a pump to a remote heat exchanger where it is cooled and then recirculated to the tube through special hoses.

Another consideration to keep in mind when choosing the type of tube housing to be used at a particular x-ray installation relates to the ***horn angle.*** The horn angle of the housing describes the relationship of the tube port to the tube high voltage receptacles, or "horns." The horn angle for a particular tube is measured in [radian] degrees. For example, a housing with a 0° horn angle will have both the high voltage receptacles and the tube port facing in the same direction; a 180° horn angle will have the high voltage receptacles pointing in the exact opposite direction of the tube port. The horn angle must be correct for the particular tube mount that will be used so that the anode will be oriented correctly, and also so the cables can be properly draped. For convenience, the standard horn angle designations are listed in Appendix C.

The final consideration with regards to tube housing type pertains to how the x-ray tube will attach to the tube supporting device. The x-ray tube can be mounted directly to the tube support by means of four bolts. These bolts are threaded through the tube mounting plate of the tube support and then directly into the tube housing.

Another way of mounting an x-ray tube is by the use of a ***trunion mount.*** In this second type of tube mount, the bolts are not threaded directly into the x-ray tube housing. Instead, ***trunion rings*** wrap around the tube housing and hold the tube in place. Most trunion ring mounts have a tension adjustment knob that can be loosened so that the technician can rotate the tube for special radiographic views.

THE HIGH VOLTAGE CABLES

The high voltage transformer unit is connected to the x-ray tube via the high voltage cables, sometimes called the high tension cables. The high voltage cable ends are terminated with a type of connector called a ***federal connector*** and plug directly into the transformer and x-ray tube (Figure 13). The cable ends are coated with a petroleum-based, dielectric (insulating) compound at the x-ray tube end, or special insulating oil is added to the high voltage transformer wells to prevent arcing.

Two important components of the federal connector are the ***grounding compression washer*** and the ***rubber gasket.*** The grounding washer must be used in the x-ray tube and transformer receptacles to ensure that the high voltage cables are safely grounded. The rubber gasket is needed to provide the proper spacing in the high voltage receptacle and to seal out moisture.

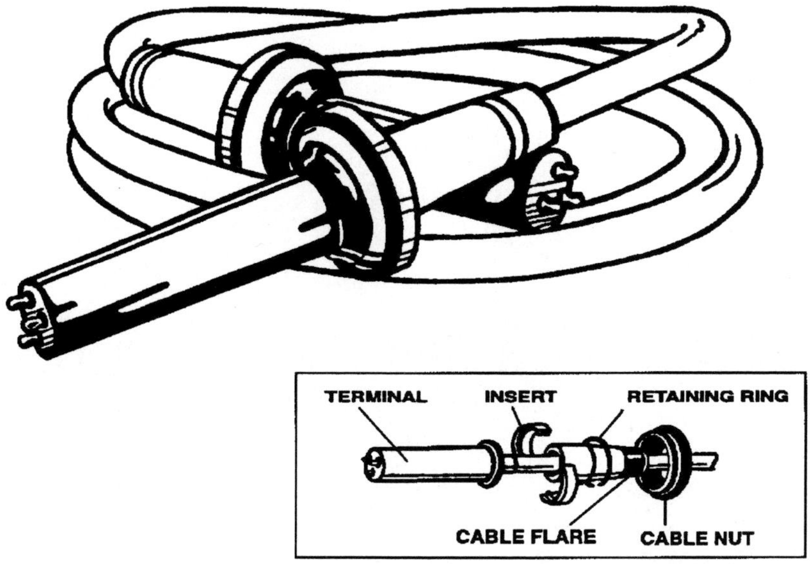

Figure 13. High voltage cable with federal connector.

Because of their construction, the high tension cables add a specific amount of capacitance to the high voltage circuit. The amount of capacitance is determined by the cable construction and also by the length of the cables. The increase in capacitance has a filtering effect on the high voltage waveform and must, therefore, be accounted for during calibration.

Chapter IV

THE GENERAL RADIOGRAPHIC ROOM

The general radiographic room has been in existence for one hundred years and is the most common type of room used today for diagnostic x-ray examinations. The same devices that were used in those early rooms are still regularly being employed in today's hospitals and clinics. Again, the "basics" of x-ray have not change since its inception!

Besides the actual x-ray generating equipment described earlier, other devices are needed in the general radiographic room in order to obtain a good diagnostic quality radiograph. A ***tube support*** is needed to position and hold the x-ray tube firmly in place during the x-ray exposure. An ***image receptor*** is needed to convert the radiation energy that is exiting from the patient to a permanent visible image. A special ***radiographic table*** is needed to properly position the patient and image receptor for the examination.

Finally, a ***collimator*** required in all radiographic rooms. The collimator is used to reduce the size of the radiation field to the area of interest, thereby reducing scatter radiation. Scatter radiation is not only harmful to the patient and hospital staff, but also causes a reduction in image quality (see Chapter VIII). Collimators also provide a visual light field that accurately defines the x-ray field.

Some radiographic rooms contain just an x-ray tube mounted on a vertical tubestand and a wall receptor. This specialized type of radiographic room is used for chest studies only, and is called a ***chest room.*** The x-ray tube and receptor can be physically attached to a single supporting column to form an ***integrated tubestand,*** or a separate tubestand and wall unit can be mechanically connected in order to operate together as an integrated system called a ***chest unit.*** If the chest unit uti-

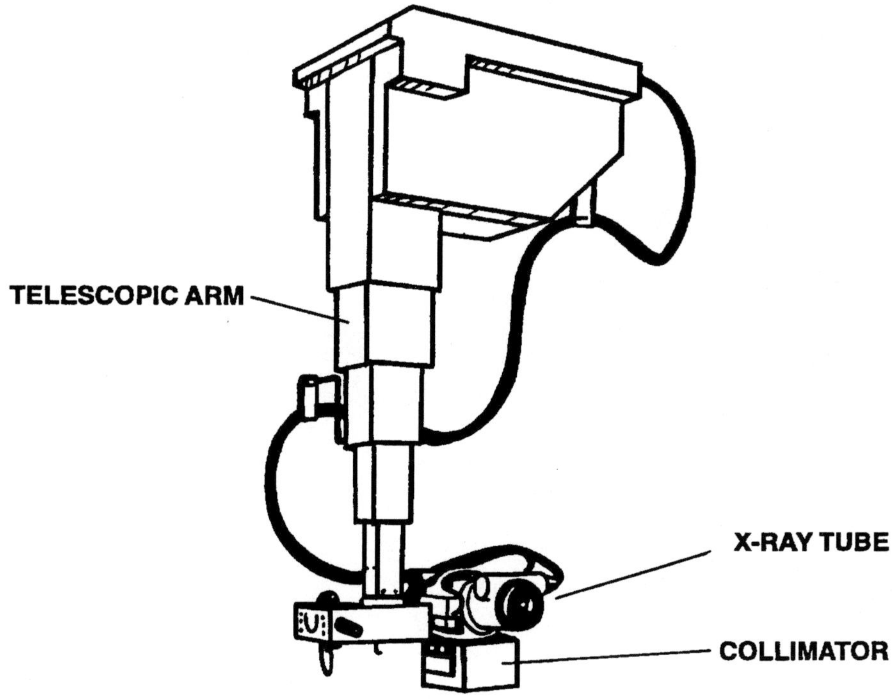

Figure 14. The overhead tubecrane. Used for positioning the x-ray tube.

lizes an automatic film transport system that is connected to a film processor, it is called an ***automatic chest unit.***

THE TUBE SUPPORT

The function of the tube support is to facilitate positioning of the x-ray tube. The two tube supports used in a radiographic room are the ***overhead tubecrane*** (Figure 14) and the ***tubestand*** (Figure 15). The overhead tubecrane is mounted to rails that attach directly to ceiling supports. The tubecrane moves along the rails to position the x-ray tube for the desired anatomical view. A tubestand is usually mounted on a track that runs along the floor and one that runs along the ceiling, but can also be mounted from the floor to a track mounted on the wall. Some tubestands are mounted directly to the x-ray table.

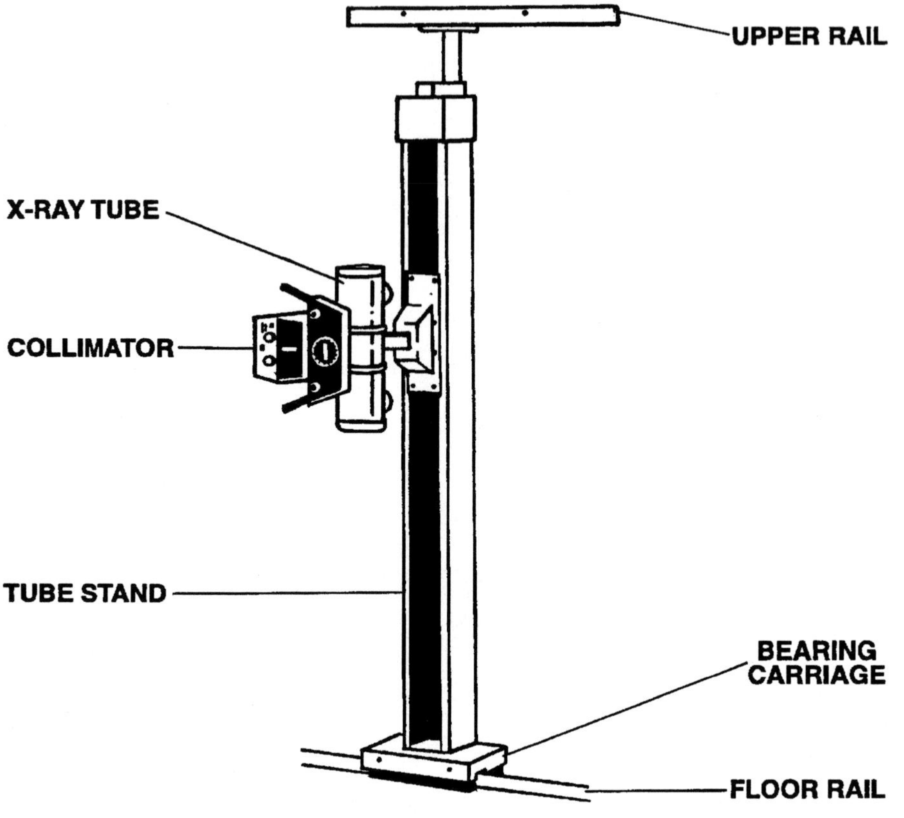

Figure 15. The tubestand. Also used for positioning the x-ray tube.

An ***integrated tubestand*** (Figure 16) was designed for installations where the size of the room severely limits the use of a separate tube support and image receptor. With the integrated tubestand, the tube support and the receptor are both attached to a single support column. These integrated units are easy to install, occupy very little space in the room, and can provide most of the radiographic views required for x-ray examinations.

Regardless of the type of tube support, its main function is to hold the x-ray tube at the particular angle, height, and rotation that is needed for the x-ray exam being performed. Most tube supports use electromagnetic locks to hold the tube in the desired position. The most basic tube support designs utilize friction locks.

The tube support must move easily in all directions. However, because of the weight of the x-ray tube, all tube supports require a

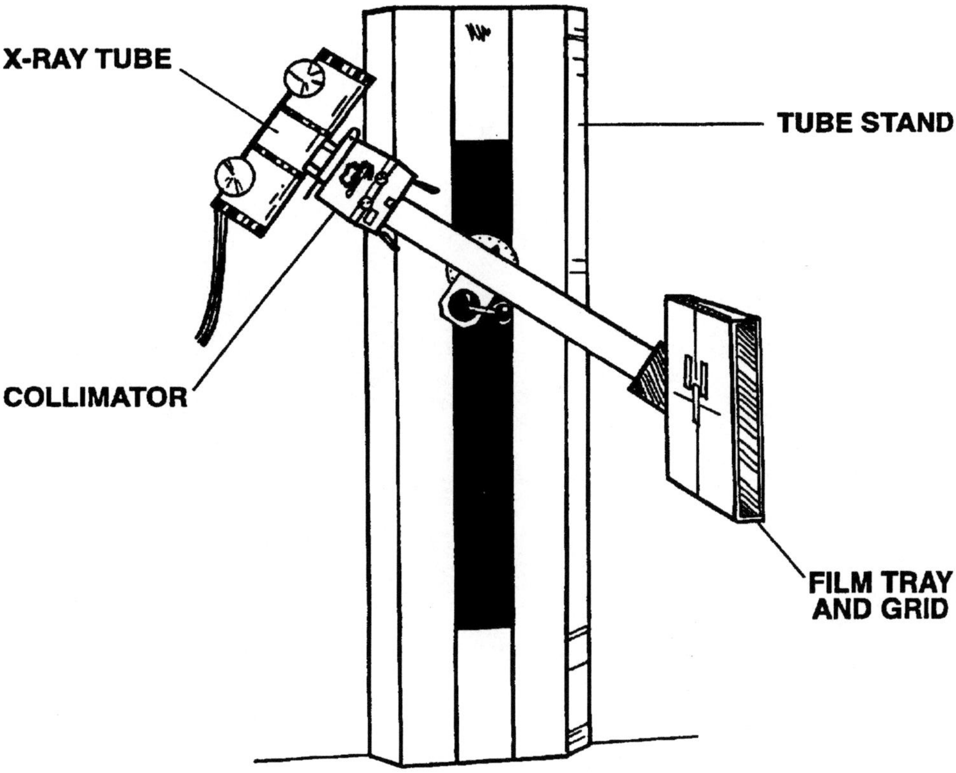

Figure 16. The integrated tubestand. The integrated tubestand positions the x-ray tube and image receptor, as well.

counterbalancing system to accomplish smooth, effortless movement in the vertical direction. The counterbalancing system includes pulleys, cables, and either counterweights or a spring tension (i.e., counterpoise) mechanism. Some tube supports can provide motorized movement, as in many integrated chest units. Other features that may also be included in a tube support are: an SID indicator, a tube angulation/rotation indicator, and a digital numerical display.

THE IMAGE RECEPTOR

The image receptor in a radiographic room is the film and film cassette. When service engineers discuss image receptors, however, they usually include the entire assembly that houses the image receptor.

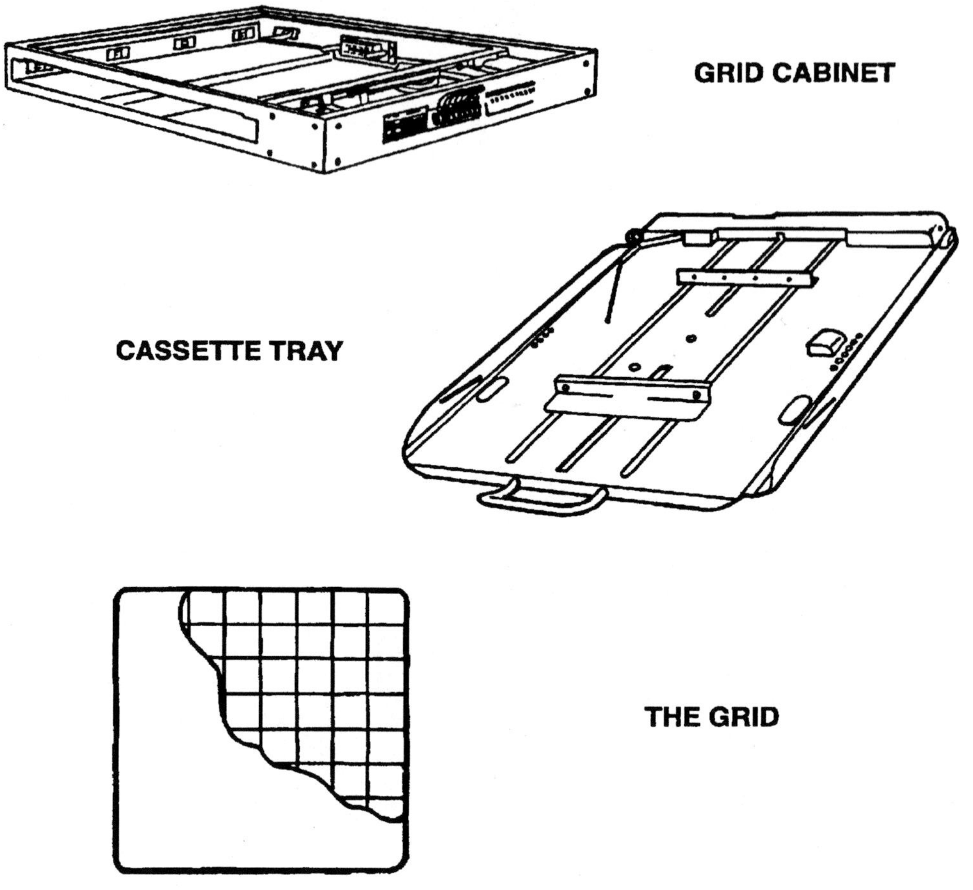

Figure 17. The image receptor. The grid and cassette tray are located within the grid cabinet.

This assembly can consist of a simple ***cassette holder,*** a ***grid cabinet,*** or a ***bucky assembly.*** In addition, the ***x-ray grid*** and the ***cassette tray*** are also considered part of the image receptor (Figure 17). Image receptors are located either on the wall for erect studies, or located in the x-ray table for studies where the patient must be in the prone position.

All receptors have a slot that allows for the placement of an x-ray grid. Grids are constructed of lead strips separated by an interspace material (i.e., wood or aluminum), and when positioned in front of the film, allows only the primary x-ray beam to pass through. The grid is used to reduce scatter radiation and thereby greatly improves image quality (see Chapter VIII).

Grids can be removable, or permanently mounted in the receptor enclosure. A stationary grid is positioned in front of the cassette tray in a grid cabinet. If the grid is a reciprocating (oscillating) type, the grid cabinet is then termed a bucky.[7] A reciprocating grid provides more scatter reduction than a stationary type and, therefore, has a greater effect on image quality. If an optional AEC device is used, it is also placed in the grid cabinet or bucky assembly.

A cassette holder assembly is very basic in design and simply consists of an upper and lower jaw that locks the cassette into position. The cassette holder assembly is usually mounted directly to the wall and can move vertically along a track. A slot is also provided in front of the cassette holder to accommodate a removable stationary grid.

THE RADIOGRAPHIC TABLE

A ***radiographic table*** (Figure 18) has several features that distinguish it from ordinary tables. All radiographic tables must have a tabletop made of a special material that will allow x-ray radiation to pass through unimpeded (technically speaking, a tabletop must have a high transmission and low absorption characteristics). The material must also have a uniform density and must be free of defects that could cause artifacts on the film. The tabletop must also be strong enough to support larger patients of 300 lbs. or more! Wood and carbon fiber materials are commonly used in table tops.

Another feature of radiographic x-ray tables is that they must contain an image receptor. Image receptors are located directly under the tabletop and usually slide in a track along the length of the table.

Starting from the most basic radiographic tables described above, optional features can be added to improve the functional capabilities of the table. These include: a hi-lo option for mechanically raising and lowering the table height, a floating tabletop option which allows the tabletop to move freely in all directions, and a tilting option that mechanically tilts the table 90 degrees.

---

7. The bucky was originally called the Potter-Bucky Diaphragm, named after Dr. Hollis E. Potter and Dr. Gustav Bucky who invented it in 1920. That name was shortened for convenience and today the "bucky" is the accepted name for that device.

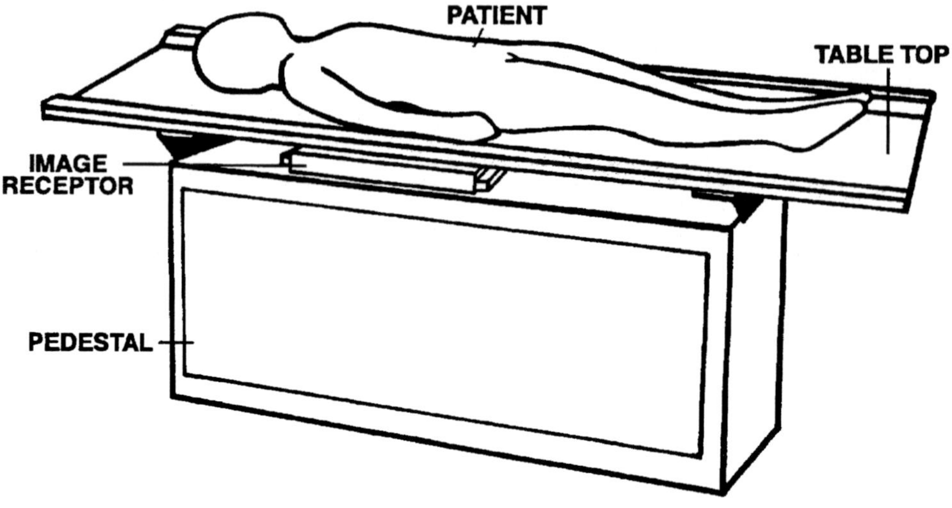

Figure 18. The radiographic table. The image receptor and table top move to facilitate positioning.

THE COLLIMATOR

The other important device in a general radiographic room is the collimator. The collimator is a beam limiting device with adjustable shutters to vary the x-ray field size. It limits the beam to the area of interest, and in doing so, significantly reduces scatter radiation and more sharply defines the edges of the x-ray field.

Because the collimator can vary the x-ray field size, it has replaced the use of ***cones*** and ***diaphragms*** in radiographic applications and is the most common beam limiting device used today. Cones were commonly used in early radiographic rooms but are now rarely used because of their fixed field size. Diaphragms are also a fixed field size, beam-limiting device, but are mainly used in mammography units.

The collimator basically consists of two sets of lead shutters positioned in such a way as to form a rectangular field (see Figure 19). ***Collimator fingers*** are attached to the upper set of shutter blades and function to eliminate off-focus or ***stem radiation.*** Another type of collimator uses an adjustable circular shutter (or diaphragm) and is normally used in conjunction with an image intensifier.

A second important function of a collimator is to provide a light field that precisely represents the x-ray field. The light field aids the x-ray

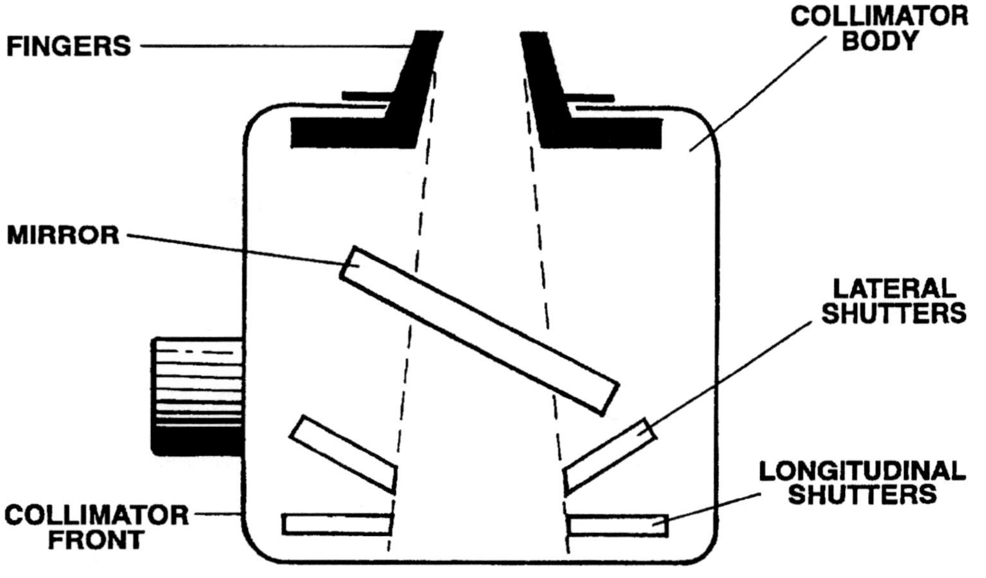

Figure 19. The collimator. The fingers are needed to remove stem radiation. The mirror adds additional filtration to the x-ray beam.

technician in positioning the patient. A centering light is also supplied by the collimator for aligning the tube with the image receptor. To produce the light field, a lamp power supply, a special lamp, a mirror assembly, and lamp timer circuit are required. These items are contained within the housing of most collimators, although the lamp timer circuitry may be located in a collimator control cabinet. All collimators also provide a slot into which additional filters may be added to modify the x-ray beam.

Shutter Operation

Collimators are designed for two modes of operation: manual or automatic. ***Manual collimators*** are more basic in design, requiring no additional circuitry other than what was described above. To adjust the field size, the x-ray technician manually turns the ***transverse*** and ***lateral shutter knobs*** to the desired field size. On the other hand, ***automatic collimators*** contain the additional circuitry required to drive the shutters automatically to the correct field size once a film cassette is inserted. This additional circuitry consists of DC motors which actually open and close the shutters, as well as some type of control logic to tell the motors when to drive and for what duration. This automatic mode of operation is sometimes referred to as ***PBL*** (Positive Beam Limitation) by some manufactures.

Also, all automatic collimators have an ***exposure interlock circuit*** which is designed to inhibit exposures if certain criteria are not met (i.e., film must be present, SID must be correct). A ***semiautomatic collimator,*** also available from most manufacturers, employs all the logic control of an automatic collimator, but without the drive motors. The shutter knobs are turned manually until a "ready" light appears on the collimator.

In the event of a failure, all automatic collimators provide a ***bypass switch*** located either on the collimator itself or in the remote collimator control panel. This switch is used to bypass the automatic mode of operation. By turning this switch, the operator switches the collimator from automatic mode to manual mode. The collimator can be used in this mode until the service technician arrives to repair the automatic circuitry.

Chapter V

SAFETY IN X-RAY SERVICING

There are several potential hazards associated with the installation and servicing of radiographic equipment. The service engineer must be aware of these "hazards" in order to safely work in this field.

While working on radiographic equipment, the service engineer will encounter ***mechanical hazards, electrical hazards,*** and ***x-ray radiation hazards.*** In addition, most radiographic equipment is installed in a medical environment where very sick patients are routinely examined. Consequently there also exists a potential risk of infection from the presence of ***biohazards.*** With an acute awareness of these potential hazards, the service engineer can make intelligent choices that will greatly reduce (or totally eliminate) his or her risks while servicing radiographic equipment.

This chapter will review the common safety hazards that are encountered in the x-ray servicing field. Much of the information in this chapter is drawn from the experience of the author and, therefore, will be presented with a "hands-on" approach.

MECHANICAL HAZARDS

Many of the mechanical devices in a radiographic room are potentially hazardous if not handled properly. The most obvious problem with equipment handling is related to the physical size and weight of x-ray equipment. As stated in the introduction, most x-ray equipment is large and rather awkward to handle. Because of the excessive weight of the individual components of an x-ray system (an x-ray tube weighs on average about 40–50 lb.), *proper precaution must be exercised when moving*

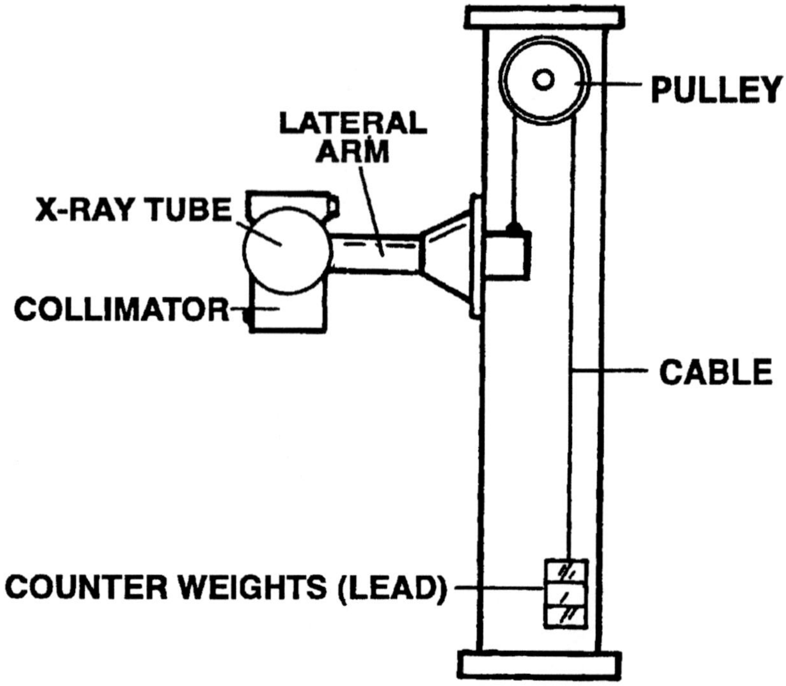

Figure 20. The tubestand counterbalancing system.

or lifting equipment. In fact, many of the duties in x-ray servicing should only be performed with the aid of a second engineer.

The ***tubestand*** is designed to hold the x-ray tube in a number of different positions. The main safety concern with a tubestand has to do with the ***counterbalancing mechanism.*** As stated above, an x-ray tube, alone, can weigh nearly 50 lbs. In order to provide smooth travel in the vertical direction, the tube must be counterbalanced with a system of counterweights, pulleys, and cables (Figure 20). The counterweight system must be inspected on a regular basis (see Chapter VI) to ensure safe operation. Also, the counterweight assembly must be secured (with blocks of wood or tied down with heavy rope) before any major service is performed on a tubestand.

The overhead ***tubecrane*** performs the same functions as a tubestand (i.e., tube positioning), but uses a different system for counterbalancing. Instead of using counterweights, tubecranes have a counterpoise system that employs ***spring tension*** for counterbalancing (Figure 21). A counterpoise provides smooth vertical motion and is very reliable.

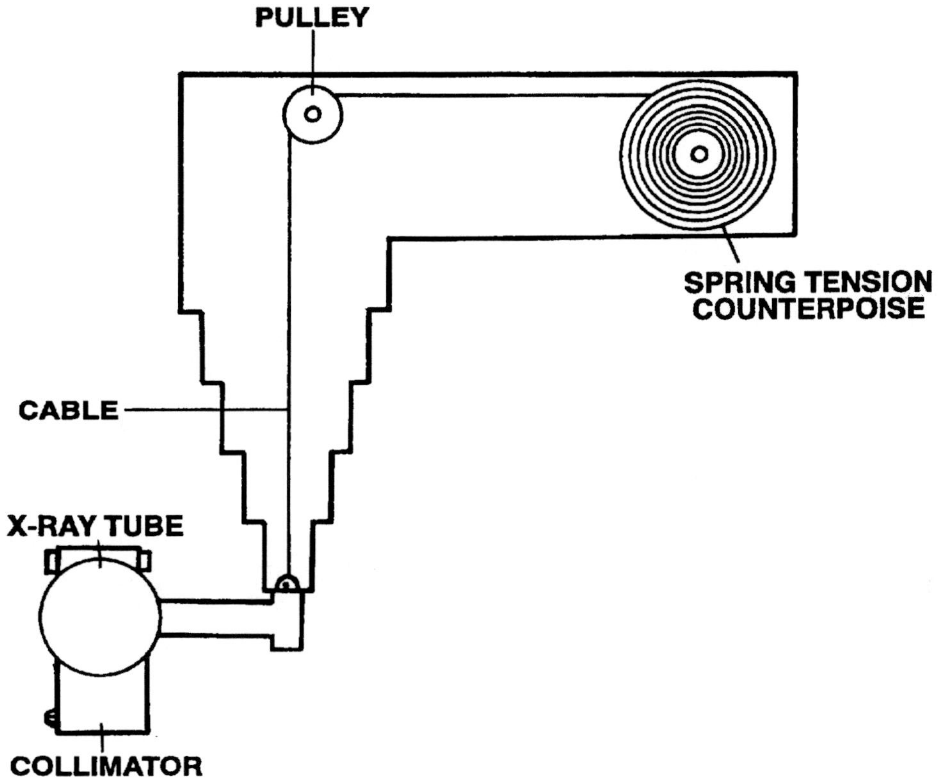

Figure 21. The tubecrane counterpoise.

Most manufacturers provide a safety locking mechanism that must be engaged when major service is needed. Again, the cables and spring tensioner should be inspected on a regular basis and caution should be used when working on this device.

The ***x-ray table*** performs two important functions: it holds the image receptor in place, and it facilitates positioning of the patient during an x-ray examination. Since most radiographic tables must support patient weights as much as 300 lbs., they must be constructed with metals of sufficient strength. As a result, these tables are usually very heavy, often requiring special "table jacks" for lifting and moving the tables during the installation. Usually though, once the table has been mounted in place, the service engineer will rarely have to move it.

Another safety hazard involved with x-ray tables has to do with the ***vertical drive system*** (if provided). The vertical drive system consists of a drive motor mechanically connected to a worm gear or chain drive system that raises and lowers the table. The motor is activated by a footswitch (or handswitch) and will drive the table up or down until it reaches a limit switch. The limit switch will then disconnect power to the motor. As a rule, power should be turned off when servicing this type of x-ray table, especially when the service engineer is working beneath the table.

If the engineer must have the table power "on" when servicing the table, the footswitch (or handswitch) for the vertical drive must be disconnected. This may involve removing the wires from the switch or from the terminal block in the table base. Most tables, however, provide plug connections for these switches that are easily disconnected. Although it takes a little extra time to perform, disabling the drive switch is easily done and well worth the effort. It is not uncommon to hear of situations where the service person was working inside a table and then accidentally leaned on the vertical drive switch and the table began to drive down on the engineer.[8] By disconnecting the drive switches, such accidents can easily be avoided.

A ***wall receptor*** or ***erect bucky*** is a device in the x-ray room that allows the technician to take exposures of patients while they are standing erect. The wallstand supports a bucky (or grid cabinet) which holds the film in place and allows the use of a grid and/or AEC detector. The vertical column is bolted to the floor and the bucky moves vertically along the column to accommodate the heights of all patients. With manually operated wallstands, a counterweight assembly similar to the tubestand is used. Some wallstands are mechanically integrated to the tubestand and are driven vertically by a motor drive system.

The same precautions must be used when servicing the wall unit because of the counterweight or motor drive system. As a rule, if the entire unit has to be unbolted from the floor to be serviced, two service engineers are required.

ELECTRICAL HAZARDS

There are many potential electrical hazards in x-ray servicing field. The engineer must be familiar with the power distribution of the par-

8. This happened to the author once during an evening service call when no one was available to help.

ticular x-ray room and know which circuits are "live" before beginning to service that equipment. Also, special care must be taken when working in the high voltage circuits because of floating potentials and residual capacitances that are present in the circuits. Countless horror tales exist of servicemen being, "zapped" by high voltage because of careless practices. A careless mistake in this area can be lethal.

The electrical hazards in x-ray servicing are present in both the primary and secondary circuits of the x-ray generator, as well as in the auxiliary equipment in the room. The ***primary circuit*** begins at the incoming voltage supplied by the power company and includes all components that are connected to the primary winding of the high voltage transformer. Safety precautions must be observed when working in this area because of the potentially lethal voltages, which can be as high as 480 VAC for 3-phase systems.

The required input voltage, as specified in the service literature for the particular x-ray unit, is fed to the main breaker panel in the x-ray room, and is termed the "primary feed." The breaker panel must be located near the x-ray control and must be identified and labeled. Sometimes there are several breaker panels in the room: one panel providing dedicated power to x-ray generator power and other breaker panels for auxiliary equipment such as collimator power or table power.

Before removing any equipment covers, the breakers should always be switched off. This ensures that all circuits inside the equipment cabinet are "dead." Turning the power switch off at the generator control will not completely remove the power from inside the generator cabinet. The reason for shutting off the main breaker is to eliminate the chance of an accidental short circuit when the engineer removes or replaces the metal covers. The compact designs of today's equipment sometimes result in very little space between the metal covers and live connections. Once the equipment covers are removed, a quick check of the power feed connections at the generator with a voltmeter will confirm that no power is present. This voltage check should always be performed as a routine safety practice when servicing equipment.

With the equipment covers off, physical inspection and analysis of the components in the primary circuit can safely be performed with the main breaker switched *off.* Once the generator is inspected and the service engineer is familiar with the layout of the primary components, the main breaker can be switched on for further troubleshooting.

With the generator powered *on,* troubleshooting can safely be performed by following these important guidelines:

1. Always probe live circuits with one hand only, keeping the other hand away from the generator.
2. Never connect test equipment to a live circuit. Always turn power off first, then attach the leads of the test meter to the circuit. Many times, the "hot" test points (i.e., a voltage present) are directly adjacent to zero reference or ground points.
3. Make sure the test leads are securely attached to the component under test.
4. Use insulated tools when probing live circuits.

The secondary circuit begins in the secondary winding of the high voltage transformer and includes the high voltage cables and the x-ray tube. ***High voltage*** could be present at any point in this circuit and, therefore, extreme caution should be used when working in these areas. In addition to high voltage, a ***low voltage (high current) supply*** for the filament circuit is also present on the cathode of the secondary.

When working in the secondary circuit, attention should be paid to several areas. One hazard involves the ***residual capacitance*** associated with the high tension cables. These cables basically contain heavy wire conductors separated by an insulating material (or dielectric). This is, in fact, the definition of a capacitor. Therefore, when current is traveling through the high tension cables, they become charged.[9]

Specifically, the cables become charged to some relatively high potential during the exposure and usually are not completely discharged (depending on the exposure technique) when the exposure is terminated. *This means that the high tension cables cannot be safely handled unless they are fully discharged first!* The service engineer can discharge the cables by grabbing the insulated portion of the cable and then touching the pins of the federal connector to a ground point on the x- ray tube housing or the high voltage transformer. Once discharged, the cables can be safely inspected or connected to test equipment.

If using invasive test equipment such as a high voltage divider, the service engineer should make connections as specified in the instruction manual of the test equipment being used. The engineer should

9. This capacitance has an effect on the high voltage waveform and will be discussed in Chapter V.

also check to see that the unused wells are filled with oil when making parallel connections. Also, *never attempt an exposure unless the high voltage cables are connected to the x-ray tube or high voltage divider!* Although a common practice years ago, it is no longer necessary today because of available test equipment. In fact, newer generators will not even allow an exposure to occur if the secondary is open, unless the engineer physically disables the filament safety circuits.

The high voltage transformer unit should not have any power applied to it when it is being serviced. All internal components should be discharged before the service engineer attempts to work inside the transformer unit. A grounding rod is provided with some transformer to accomplish discharging.

All other devices in a radiographic room, such as the table, collimator, or tubestand, are usually powered by 110 VAC, single-phase. The service engineer should also use caution when working on these devices. In addition, many low voltage circuits exist (i.e., logic supplies), and although caution should be used when troubleshooting these circuits, mistakes often result in damage to the component rather than the service engineer.

RADIATION HAZARDS

Before servicing radiation producing equipment, a thorough knowledge of the nature of x-ray radiation is essential. Most people know that exposure to radiation is hazardous and should be avoided, but understanding the properties of radiation and how x-rays interact with matter will allow the service engineer to make intelligent choices that will thereby minimize his or her exposure.

Working around radiation becomes routine after a few years and many service professionals often become lax in their safety practices. The service engineer must constantly be aware of his or her surroundings and the amount of radiation that is being emitted. This point cannot be stressed enough, and must always be first and foremost on the service engineer's mind.

To properly diagnose and repair x-ray equipment, exposures must be taken. In fact, it is not uncommon for the service engineer to take as many as one hundred exposures during an x-ray calibration period. If the proper safety precautions are not taken, the service engineer could be exposed to significant amounts of radiation.

Since x-rays are invisible to the human eye and can only be detected with special test equipment, the service engineer must thoroughly understand the nature of x-ray radiation and how x-rays react with matter. With this understanding, the engineer can be aware of the amount of radiation, if any, that he or she is receiving.

The Nature of the Primary Beam

In the previous chapter, we discussed the principles of x-ray production. Here a more detailed description is provided. When the accelerated electrons collide with the target in an x-ray tube, two types of radiation are emitted: ***brems*** or ***braking radiation*** and ***characteristic radiation.*** Brems radiation is produced by the sudden deceleration of the electrons as they impact with the target and contains x-rays of many different energy levels. In fact, Brems radiation makes up most of the primary beam at the kVp's used in diagnostic x-ray.

The second type of x-ray radiation, characteristic radiation, occurs when the accelerated electron hits a target atom and causes an inner orbital electron of that atom to be ejected. This creates a hole in the inner shell of the electron which must then be filled to make the atom stable again. When this hole is filled, radiation of a distinct (characteristic) frequency for that particular atom is released. However, characteristic radiation makes up only a small percentage (10%) of the primary beam for diagnostic radiation, and is only present above 70 kVp. As a result of these two methods of x-ray production, the primary beam consists of x-rays with a wide range of frequencies (i.e., is ***polyenergetic***).

How X-rays Interact with Matter

Because x-ray radiation is a form of electromagnetic radiation, it shares many of the same properties of other types of electromagnetic radiation. For instance, x-rays travel at the speed of light and in straight lines. However, because of their high energy level, x-rays differ from all other forms of electromagnetic radiation in that they react with the matter that they come in contact with.

When x-rays interact with matter, two things can occur. The x-rays may simply pass directly through the material unaffected, or they may be absorbed into the matter and interact with the atoms in the tissue. The ability for x-rays to pass directly through matter is referred to as the ***penetrating ability.*** The penetrating ability is a function of the energy level of the x-rays and is directly related to the amount of kVp applied to the x-ray tube (see Chapter VIII). The x-rays that are absorbed into tissue will interact with the matter, resulting in radiation of lower energy exiting the tissue. This ***attenuated*** exit beam is, in fact, what forms the latent x-ray image (Chapter VIII).

When x-rays are attenuated, several interactions occur within the tissue that create a hazard for the patient, as well as, the service engineer. One type of interaction involves the ***ionization of matter.*** Ionization is the process of removing or adding an orbital electron of an atom, and thus creating an unstable ion. Ionization of living matter is known to cause chemical and biological damage (i.e., it destroys tissue). Another result of the interaction of x-rays with matter is that the high energy x-rays will liberate heat when passing through matter which can also damage living tissue.

When considering radiation safety, a third phenomenon of x-ray interaction must be discussed. Specifically, when x-rays interact with matter, two additional forms of radiation are produced, ***scatter radiation*** and ***secondary radiation.*** Scatter radiation and secondary radiation will have a detrimental effect on the patient, and are a major source of radiation exposure to the radiologist, the x-ray technician, and the service engineer. In addition, these two forms of radiation greatly affect image quality.

Scatter and secondary radiation are produced by distinct interactions with matter. The interactions that are of the greatest importance in diagnostic radiology are the ***photoelectric interactions*** and ***compton interactions.*** In photoelectric interactions, the x-ray photon is absorbed into an atom and that atom then gives off an orbital electron which creates an ion as described above. The orbital electron that was ejected becomes a high energy *photoelectron*. In addition, when the hole is filled in the atom, ***characteristic radiation*** is emitted.

Both the photoelectron and the characteristic radiation are forms of ***secondary radiation.*** Because each material (i.e., bone, soft tissue, etc.) will give off its own characteristic radiation, the photoelectric effect contributes to image contrast and, therefore, plays an important role in forming the final image.

In ***compton interactions,*** the x-ray photon collides with an outer shell electron and is *deflected off in another direction and at a lower energy level.* The deflected, lower energy photoelectrons are referred to as ***scatter radiation.*** Because the radiation scatters in unpredictable directions, *image quality is greatly reduced.* The point to emphasize regarding scatter radiation is that the x-rays are no longer just traveling in one direction (i.e., towards the film) but now travel in many directions including towards the service engineer! Because scatter radiation exposes the patient to more radiation than necessary, is hazardous to the equipment operator, and decreases image quality, an important goal in radiography is to reduce the amount of scatter radiation as much as possible (see Chapter VIII).

Radiation Measurements

Since x-rays cannot be detected by the human senses, indirect methods of detection and measurement must be employed. Most test instruments used for radiation measurements take advantage of the ionizing effects of radiation. Some of the effects of ionization are:

1. It *causes chemical changes* as with photographic film.
2. It *causes certain materials to glow,* such as phosphor.
3. It *ionizes gases.*

One method commonly used to monitor the amount of radiation received by individuals who work around radiation is by the use of a ***film badge.*** The film badge is worn by the service engineer, radiologist, and x-ray technician on a specified area of the body for a given period of time (usually one month). The badge is then sent to a testing laboratory where it is analyzed for radiation exposure. The results are documented in a monthly report and sent back to the individual.

Another method of detection is by the use of an ***ionization chamber.*** The air within the ionization chamber, or ***probe,*** becomes ionized in proportion to the amount of radiation it receives and then can be electrically measured. Most radiation physicists use this type of device to measure the output of x-ray generators and to calculate entrance doses delivered to patients. X-ray service engineers will also use the ion chamber detector for certain calibration procedures.

Radiation Measurement Units

The standard unit of measurement for radiation output of x-ray equipment is the ***Roentgen*** or ***R.*** The Roentgen is defined *as the amount of radiation needed to ionize one cubic centimeter of air to one electrostatic unit.* As a rule, x-ray service engineers will measure radiation output using "R" units when performing compliance testing and, therefore, should be familiar this unit of measurement.

When measuring radiation that is absorbed into living tissue (i.e., ***entrance dose***), other measurement units must be used. Although entrance doses can be measured in "R" units, it must be emphasized that the Roentgen is based on ionization ***free in air,*** and does not take into consideration the effects of ***backscatter radiation.*** For measurements that include backscatter radiation, there are two other units commonly used: the ***RAD*** and the ***Gray*** or ***GY.***

The ***RAD*** is a measurement of ***radiation absorbed dose*** (in tissue). One RAD equals *the absorption of 100 ergs of energy per gram of irradiated matter.* The ***Gray*** is the international unit of absorbed dose and *is equal to 1 joule of radiation energy absorbed per kilogram of tissue.* The relationship between the RAD and the Gy is: ***1 RAD = .01 Gy.***

The RAD and Gray take into consideration the backscatter factor of tissue. Since backscatter adds to the total amount of radiation absorbed in tissue, it is important to know what unit of measurement is being used when dealing with entrance doses. For example, if a measurement of entrance dose is given in "R" units, that number must be multiplied by a conversion factor of 1.21 to get the RAD equivalent.

Not all types of radiation, however, produce the same effect in tissue. Alpha particles, for example, produce much more injury to tissue than the same amount of x-ray radiation. To account for this difference, two other units of measurement were developed that take into consideration the ***Relative Biological Effectiveness (RBE)*** in tissue due to various types of radiation.

For x-ray radiation, the two units of measurement of absorbed dose are the ***REM*** and the international unit, the ***Sievert (Sv).*** One REM is equal to one RAD multiplied by a ***quality factor.*** For diagnostic x-ray radiation, the quality factor is, conveniently, equal to unity or one, and therefore, ***1 RAD = 1 REM*** (or 1 Sv = 1 Gy).

It is known that certain tissues are more susceptible to radiation damage than others. For example, the hands can receive much more radiation than the eyes or gonads, without increasing the risk factor for cancer. Most guidelines for radiation exposure, however, are stated for ***whole body*** accumulated dose. The legal limit for radiation workers is 5 REMS per year (whole body). If an x-ray service engineer's yearly accumulated dose is below this value, he or she can safely work in this field without increasing his or her risk factor for cancer.

Radiation Safety

With a thorough understanding of the nature of x-ray radiation and how it interacts with matter, the engineer can take the proper safety precautions to reduce his or her exposure. There are three things that the engineer can do to reduce (or completely eliminate) the amount of exposure to radiation he or she receives:

1. Use the ***proper shielding***
2. Maintain a maximum ***distance*** possible from the x-ray tube
3. Reduce the amount of ***time*** that x-rays are being emitted

The service engineer should always initiate exposures from behind the lead barrier of the control booth during calibrations and testing. If the engineer must be in the x-ray room while exposures are being taken, he or she must wear the proper lead shielding. This includes: a lead apron, thyroid shields, and lead gloves if placing objects in the path of the primary beam.

The collimator blades should be closed when performing x-ray calibrations and testing to eliminate any chance of exposure to the service engineer. The service engineer should never make exposures with the collimator completely removed from the x-ray tube. The collimator provides a significant amount of shielding and dangerous amounts of radiation are present without it in place. If exposures must be made without a collimator in place, ***a lead blocker must be placed on the port of the tube.***

The second factor that can greatly affect the amount of radiation exposure that the engineer will receive while servicing radiographic equipment is ***distance.*** The service engineer should always observe the ***inverse square law,*** which states that *the intensity of radiation at a given distance from a point source is inversely proportional to the square of the distance.* This means that when the engineer stands a certain distance from the x-ray tube and is receiving a known dose, increasing the distance from the tube by a factor of "two" will effectively reduce the radiation dose he or she is receiving to one-fourth!

This is one of the most important factors that can significantly reduce the exposure to the engineer. The service engineer can safely work in this field if he or she is conscious of his or her surroundings, and in particular, the proximity to the x-ray source.

The third factor that will reduce the service engineer's exposure to radiation is ***time.*** Simply put, if the engineer reduces the amount of time he or she is exposed to radiation, personal exposure will correspondingly be reduced. The engineer should avoid taking unnecessary exposures when troubleshooting x-ray problems. This obvious point is often overlooked by the service engineer who becomes "engrossed" in a complicated x-ray problem and is not conscious of the amount of exposures that he or she is taking.

Furthermore, troubleshooting should be performed using the shortest possible exposure times that will allow the engineer to make a diagnosis (this practice also extends tube life). Since exposure to radiation has an ***accumulative effect*** on tissue, each individual exposure that the engineer receives must be added to the total amount received through out his or her lifetime. For example, a person who receives a single exposure of 5 RADS suffers the same risks as a person who receives 5 exposures of 1 RAD each.

The service engineer should always wear a radiation badge when servicing equipment and should closely monitor the monthly exposure reports. Finally, the engineer should never place any part of the body in the path of the primary beam. He or she should use appropriate ***phantoms*** to check image quality. A practice used many years ago in x-ray service was to place one's hand (or body!) in the radiation field to check image quality. This practice, however, is very dangerous and is absolutely unacceptable.

BIOHAZARDS

Health hazards also exist for the service engineer working in the medical x-ray field. As with all health care professionals, the x-ray service engineer must be aware of the potential risk of infection from biological hazards when servicing x-ray equipment. Infection control has especially come to the limelight given the rise in number of ***bloodborne pathogens*** such as HIV and HBV.

The service engineer will, on occasion, be required to service equipment that is contaminated with body fluids such as vomit, urine, feces, or blood. Oftentimes the fluid itself will cause the failure in the table.[10] The point is that the danger of infection to the service engineer is very real and ever-present.

Universal precautions can be taken to prevent contact with potentially infectious materials. The equipment must be cleaned thoroughly, preferably by an in-house staff member, before the engineer begins servicing. The cleaning fluid must consist of a fairly strong disinfectant solution, usually available on the site.

In addition, the engineer must wear ***personal protective equipment*** when cleaning or servicing contaminated equipment. Rubber gloves should be worn when cleaning up spills. Goggles or face shields should be worn when larger spills are involved because of the possibility of splashing. Gowns and shoe covers should also be worn when working on heavily contaminated equipment. The service engineer should not hesitate to ask for the protective equipment because of the serious nature of infection from bloodborne pathogens.

Finally, hands should be washed thoroughly after working on patient equipment with an antibacterial soap (available at all hospitals). Also, *NEVER TAKE FOOD INTO THE X-RAY ROOM!* This common practice of field service engineers who, because of time limitations end up eating their lunch while performing repairs, should be avoided because of the risks involved.

---

10. Bodily fluids are very conductive and will "short out" components with which they come in contact.

Chapter VI

THE INSTALLATION OF RADIOGRAPHIC EQUIPMENT

The installation of radiographic equipment requires knowledge and expertise in a wide range of disciplines. Some of these areas include: knowledge of building construction; techniques for moving, aligning, and mounting (often very heavy) equipment; good electronic calibration skills; a complete understanding of x-ray imaging principles; and knowledge of film and film processing techniques. The installer (i.e., the x-ray service engineer) must be well-rounded in all of these areas in order to properly install a radiographic room successfully.

In addition, when the installation is completed, the room must look aesthetically pleasing. A good room layout design is needed to accomplish this goal. Great care must be taken when designing a room layout to ensure that the radiographic equipment will fit in nicely with the rest of the fixtures in the room. A radiographic room that is attractively laid out will present an image that reflects positively on the equipment installer (and his or her organization).

While the engineer is performing the installation, **all wiring and** *special modifications should be documented* in the service manual. It is extremely frustrating for a service engineer to attempt servicing x-ray equipment that has been modified in some manner, when there is no appropriate documentation of the modification at the site. Equipment which has circuits that have been modified (or completely disabled) should be noted in the service literature and the actual modification should be clearly spelled out on the schematic. Also, all pertinent blueprints and schematics that show the exact power distribution and cable runs should be stored with the equipment manuals. Without this information, troubleshooting can take much longer than necessary.

Once the installation has been completed, the room is then carefully tested by a senior service person so that any items that may have been overlooked or that do not meet specification can be corrected at this time. When the room is ready for operation, the engineer may be asked to demonstrate the safe and proper operation of the equipment (i.e., user inservicing), although this task is usually performed by a clinical application specialist. After the room has been in operation for a month or so, the engineer should revisit the site to check on any problems that might have been encountered during the "break-in" period.

The installation of any radiographic room, whether a new installation or a reinstallation of used equipment, can be divided into several stages. These are:

1. Site Planning
2. Room Preparation and Prestaging
3. Equipment Delivery
4. Equipment Alignment and Mounting
5. Calibration
6. Final testing
7. Inservice of Staff
8. Follow Up

This chapter will review important guidelines for the installation of a ***general radiographic room*** covering all of the stages listed above. These guidelines are written in a somewhat generalized form and therefore can apply to most "brands" of x-ray equipment. After reading this chapter, the service engineer should have a good understanding of what is to be accomplished during an installation. However, when actually installing x-ray equipment, the specific installation instructions of the particular brand of equipment should be followed. *The manufacturer's guidelines are written specifically for their particular equipment and should be followed whenever possible.*

SITE PLANNING

The radiographic room should be designed so that the equipment will perform to the customer's expectations as specified in the sales agreement. To accomplish this goal, site planning meetings are arranged between the customer and the equipment dealer. A service engineer must be present at these meetings to provide any technical information that is needed.

Several factors influence the location of the specific x-ray devices in the room. The room layout is mainly dependent on the size and weight of the equipment and the actual room dimensions. In some cases, the cables that interconnect between the various x-ray devices in the room are restricted in length and will also limit the number of layout possibilities. Other considerations in room layout are: patient accessibility (i.e., with stretchers or wheelchairs); service accessibility; convenience for radiologists and technicians when performing studies; and proximity to the darkroom.

All of these points can be addressed with a good room layout drawing or ***blueprint.*** The blueprint is a scaled-down drawing of the actual room and is used by the contractor during the room preparation phase. Usually, the hospital staff, along with representatives from the equipment dealer (i.e. engineer, architect, salesperson), work together to produce the best possible room layout.

If the equipment is being installed in an existing x-ray room, the blueprint for that room will aid in the decision as to where the new equipment will be located. By using the existing framework and conduit, the cost of the new installation is greatly reduced. Also, the proper amount of lead shielding should have been installed in the walls and doors. This saves much time during the room preparation phase and also reduces the overall cost of the installation. The usual practice is to follow a similar layout of the existing room. This can be easily accomplished with only minor compromises.

If the equipment is being installed at a new site, more preparation is needed and greater cost will be involved. On the positive side, there is more flexibility in the layout design with new construction. The service engineer can dictate exactly where the conduit should run, where junction boxes should be mounted, and the number and size of wires that should be pulled through the conduit. Also, if any walls, ceilings, or floors need to be reinforced to support the new equipment, it can easily be done with new construction.

The specifications for all building materials (i.e., supporting framework, lead shielding, electrical wire) and any applicable building code requirements are usually listed on the blueprint. This information is also available from various sources including the equipment manufacturer, the building inspector, the director of plant operations in the medical facility, as well as from the contractor who is performing the work. However, information on the radiographic equipment specifications (i.e., equipment size and weight, power requirements, environmental operating conditions) is available directly from the equipment manufacturer and is usually included with the sales and/or service literature.

Usually the darkroom is also included in the layout design. The darkroom must be relatively close to the x-ray room and should be easily accessible. All plumbing requirements, as well as the ventilation specifications, should be documented on the blueprint. If chemical resistant floors or walls are used, this should also be specified. If the darkroom is adjacent to the x-ray room, proper thickness of lead shielding should be documented, as well.

There are, however, other important details that must be considered when designing the layout of an x-ray room. Great care should be taken to ensure that *the equipment is installed with future servicing in mind.* Each piece of equipment in the room should be installed so that it will be easily accessible for servicing. Furthermore, the equipment cables for each device in the room should be of sufficient length to allow for easy movement. Many times, equipment which can only be serviced through the rear panel will be positioned directly against a wall for aesthetic reasons. In this case, the service engineer must have the ability to move the unit away from the wall to gain access to the rear panel.

Also, the control console, if possible, should be located in the vicinity of the x-ray power cabinet (for high frequency units) or the high voltage transformer (for single and three-phase units). This is done so that one service engineer *alone* can conveniently perform troubleshooting and calibration procedures on the system. The service engineer must be able to connect test equipment to the generator section of the x-ray unit and view measurements and signal waveforms while taking exposures at the x-ray control.

If the high voltage transformer, for example, is located in a room of the facility that is different from where the x-ray control is located (or on another floor), the service engineer will not be able to check specific test points or make adjustments while taking exposures. In this less than ideal situation, the engineer would have to make an exposure at the control and then run to the room where the generating equipment is located to view stored signals and make adjustments.

One experience with this scenario will convince the engineer of the foolishness of this type of equipment layout. In fact, it is almost impossible to perform proper servicing with one engineer in this situation. Here the service organization must send two engineers to perform tasks that are normally performed by one engineer.[11]

The proper room layout can only be accomplished if an experienced x-ray service engineer is involved in the entire installation project from initial planning to completion. The experienced engineer will be able to decide the best layout of the equipment so that all parts are accessible for repair and that the proper room lighting is available to aid in servicing.

Many times because of size limitations of the room, compromises must be made regarding the room layout. These compromises, however, should never limit equipment accessibility for servicing. Also, an x-ray service engineer should be present on site during the entire installation to ensure that the room is being properly installed and is service friendly.

Once the plans have been drawn and accepted by all the parties involved, a ***preliminary time schedule*** for the installation is made. This time schedule shows the projected "time frame" for room preparation, equipment delivery, and room completion. Certain circumstances, such as construction problems, or damaged or missing equipment, will occasionally arise and the preliminary schedule will have to be modified. However, it is in the best interest of all parties concerned to stick to the preliminary schedule if at all possible.

ROOM PREPARATION

With all matters detailed on the blueprint, the contractor can proceed with the room preparation. It is at this time that the ceiling and floor supports are installed (or checked, if previously installed), the lead shielding is installed (or checked), all electrical conduits are laid out, and the power is brought to the main breaker panel in the room (usually within reach of the operator control booth) Also, all lighting and ventilation systems will be installed. Generally, the room is painted and

11. The author has serviced x-ray equipment where the high voltage transformer was located in a remote hallway, far away from the x-ray control. Consequently, an intercom system had to be installed so that the engineers could communicate with each other.

flooring is laid at this time and then later touched up at the completion of the installation.

During the room preparation the x-ray service engineer should visit the site to ensure that everything is going as planned and to advise if changes or problems have been encountered. Also, all inspections should be made at this time by the appropriate officials (i.e., building inspectors and radiation physicists).

PRESTAGING EQUIPMENT

While the room preparation phase is taking place, the x-ray service engineer can uncrate and inventory the new equipment, which is usually temporarily stored at the equipment dealer's warehouse. In addition, many of the x-ray devices can be configured for their intended operation while at the warehouse and then tested ahead of time. The process of setting up (or configuring) and testing equipment is called ***Prestaging.*** Ideally, prestaging should be performed before every installation if the manpower is available.

Prestaging will absolutely reduce the actual time of the installation! Many important tasks can be completed during prestaging phase that will reduce the actual installation time at the site. These include:

1. Uncrating and equipment inspection
2. Inventory check
3. Equipment configuration
4. Power-up testing

Uncrating Equipment

When equipment is shipped from the manufacturer, it is packaged in large, reinforced crates to ensure safe shipping. This special packaging is no longer necessary once the equipment is safely in the dealer's warehouse. By uncrating the equipment, the service engineer can then carefully inspect each item for damage. Furthermore, the equipment can also be transported more easily to the site without the bulky crates. Usually, there is not enough space available at the installation site for these crates anyway.

Before the equipment is uncrated, a visual inspection of the crate itself should be performed. A damaged crate indicates mishandling during shipping and possible damage to equipment. Any damaged crates should have been documented for insurance reasons at the time they were delivered.

Once the equipment is out of the crate, the engineer should look for any signs of physical damage due to faulty manufacturing or shipping. These signs include dents or scratches on the equipment surfaces and any damaged or missing hardware. In addition, the condition of the paint on each device should be checked for uniformity and the correct color must be verified. All moving parts for each device should be checked for smooth operation and the bearings and bearing tracts also inspected. The hardware and brackets for each piece of equipment should be inspected and tightened, if necessary, at this time.

Inventory

After each device has been inspected for physical damage, an inventory check should be performed. Each item should be checked against the shipping list, and then compared to the original purchase order. The inventory check should include not only equipment but also accessory items such as cables and user manuals. The engineer should check that all cable lengths are correct, as some cables are manufactured specifically for the particular job and are not a stock item. If a new cable has to be ordered, for instance, it could delay the installation several weeks. Finally, during the inventory check, all the serial numbers should be recorded and documented on the installation report and on the equipment registration forms.

Equipment Configuration

If every piece of equipment is accounted for and in good physical condition, the service engineer can begin to configure the equipment for its specific use. Configuration is a process of programming the equipment both mechanically and electronically for its intended use at the particular facility. Mechanical configuration involves installing the correct hardware so the equipment will operate properly as specified in the room layout. Specifically, wall receptors and collimators must be configured for left hand or right hand operation, the proper x-ray tube mounting hardware must be installed for the particular tube support used, a bucky or grid cabinet must be installed in the x-ray table, and the rails for the overhead tubecrane (if applicable) must be cut to length.

Electronic configuration includes any special wiring that must be installed on each device, the setting of dip switches on circuit boards, and software programming. Any special cables that are needed should also be "made up" (i.e., cut to length and terminated) at this time.

X-ray Generator Configuration

The specific functions that are electronically programmed in the x-ray generator include: the type of incoming power, the number and type of x-ray tubes being used, the number of image receptors, the presence of the AEC option, the type of rotor controller being used, and any special radiographic applications that will be installed. Configuration is usually accomplished by installing jumpers, or by programming dip switches on the circuit boards inside the generator control.

With x-ray equipment that is microprocessor controlled, much of the configuration is done through the software. Specifically, the service engineer will enter into a "service mode" at the x-ray console and information is then entered into memory through the soft keys on the console. The data is stored in nonvolatile ram called EPROM's and will remain there until the engineer re-enters the service mode.

Line Matching

Line matching must be performed on all generators before power can be applied. If the generator is not configured for the exact incoming line voltage, the proper calibration can never be achieved. Moreover, the internal components of the x-ray generator could also be damaged with an incorrect line setting.

Configuring the generator to match the incoming line (i.e., line matching) is accomplished by adjusting a wire tap on the line auto transformer inside the x-ray generator. Most autotransformers can accommodate a wide range of input voltages. If the required tap for a given input voltage is not available and no other source of power is available in the building, then a line matching transformer must be used.

X-ray Tube Protection Programming

X-ray tube protection programming is a very important step in the generator configuration process. The purpose of this programming is to ensure that the tube will operate safely within its designed limits so that no possible damage can occur to the tube itself. Once the generator is correctly programmed, the operator will be prevented from selecting an exposure technique that exceeds the tube rating specification for that particular x-ray tube.

In order for the "tube protect" circuitry to function properly, the exact characteristics of the particular x-ray tube must be known. The engineer must refer to the ***x-ray tube rating charts*** for information about the particular tube being used. These charts are supplied by the manufacturer of the x-ray tube and are usually shipped with the x-ray tube. The tube rating charts contain all the specifications of the x-ray tube insert in graphic form, including anode heating and cooling charts, filament emission charts, and exposure rating charts. Also, information about focal spot size and maximum continuous filament current is provided.

Additional information regarding the tube housing is also provided with the tube rating charts. This includes the type and power requirements of the stator, the maximum allowable kilovoltages for the particular housing type, and the thermal ratings for the particular tube housing.

The engineer must be comfortable with using the tube rating charts and must know how to interpret the data they contain. If the charts are not interpreted properly, the x-ray tube may not operate at its full capacity, or even worse, may operate over its maximum rated capacity. In the latter case, x-ray tube damage will most often result. To aid the service engineer, a sample tube rating chart is provided in Appendix B along with guidelines on how to interpret them properly.

Once the engineer reviews the charts and properly interprets the tube data, x-ray tube protection programming can begin. Programming is accomplished in several different ways since the tube protection circuits differ in the level of sophistication among manufacturers. However, the most basic designs require the placing of jumpers, potentiometer adjustments, or the setting of dip switches to activate a tube limit when the maximum allowable technique is selected.

The latest generator designs, on the other hand, utilize preprogrammed software that is installed at the factory and shipped with the generator. This software, usually stored in a PROM chip, contains data for several standard x-ray tubes that are available from various manufacturers. The engineer will simply select a tube from the list by means of dip switches or through soft keys at the console. For x-ray tubes that are not commonly used, the manufacturer of the generator can "special-order" the correct software.

Other x-ray generators provide the ability for the engineer to program the actual tube data (read from the tube rating chart) directly into memory through the console keys. Any type of x-ray tube can be programmed into these microprocessor controlled generators. In this case, all the pertinent tube information must be manually entered into the x-ray control memory. To accomplish this, data points are read directly from the tube charts and entered through the "soft keys" into the memory in a special programming mode. These data, points are used to create an arithmetical curve that will accurately represent the safe operating limits of the x-ray tube. This procedure takes time to complete but should not be rushed since this information must be accurate for the tube protect circuits to operate correctly.

Power-Up Testing

Power-up testing should be performed as a final check to verify equipment condition. Most of the time, testing is done to each piece of equipment separately, that is, not yet connected as a system. Before applying power to the unit, any wires, plugs, or connectors that may have become dislodged during shipping should be checked and reconnected. In addition, all circuit boards should be reseated. Some x-ray generators cannot be powered-up with an open secondary circuit (i.e., without an x-ray tube connected). In this case, the manufacturer's procedure for safe power-up should be followed.

When performing power-up testing on an x-ray generator, all the panel lamps and indicators should be checked for proper functioning, and all the switch functions should be tested. Microprocessor-controlled units usually have built-in, power-up diagnostic checks that are automatically performed when the unit is first turned on. These diagnostic programs will send an error code to the console if any problems are encountered during the "boot up" period. The error code will point directly to the area of the failure, thus saving much valuable time during the installation. In addition, many generator functions can be tested via the keypad at the console during this time.

Once powered-up, the unit should be left on for a period of time to allow the components to warm up and reach their normal operating temperature. Defective components will often fail at this time and replacement parts can then be ordered before the equipment is delivered to the site. If the generator passes the power-up testing and there is sufficient time, the x-ray tube should be inspected and then connected to the transformer for a preliminary calibration (refer to the calibration section of this chapter).

X-ray Tube Inspection

Before an x-ray tube can be installed into the system, it must first be physically inspected. The tube housing should be examined for dents or for any signs of oil leakage that could indicate a defective seal. The engineer should then check for the presence of air bubbles inside the housing. This is done by looking directly into the port of the tube while gently rocking it, first by lifting the cathode end and then lifting the anode end. If any air bubbles are observed, the tube is defective and must be sent back to the tube manufacturer for reloading.

The engineer should next perform several checks on the anode. These checks include: a visual inspection of the anode, measurements of the stator resistances, and a checkout of anode rotation.[12] The service engineer should inspect the anode surface by shining a, flashlight directly into the port of the x-ray tube. The anode surface should appear smooth and uniform. Any obvious signs of defects such as cracks, pits, or warping should be documented and the tube should be returned for repair.

The engineer should then measure the resistances of each stator winding and compare the actual readings with the nominal readings listed in the tube housing data sheet that was shipped along with the

---

12. A more detailed starter verification procedure is performed during the calibration phase. At this time a check for correct rotation is adequate.

tube housing. Once confirmed, the three stator supply leads (main, phase, and neutral) can be connected to the x-ray tube and rotor controller.

Proper anode rotation is verified by applying a starting voltage to the windings by briefly tapping the "prep" switch.[13] A flashlight is again used to determine that the anode is rotating smoothly and in the direction specified in the tube information package. Also, the engineer should listen for quiet operation while the anode is rotating and note the ***coasting time*** of the anode when the voltage is removed (i.e., when the prep switch is released). The anode should coast for several minutes (minimum) before coming to a complete stop.

Next, the filaments should be examined. Before connecting the high voltage cables, however, the continuity of the filament wires should be checked. One way to measure the filament wire resistance is to insert a test connector into the cathode well.[14] The resistance readings are taken from the small focus lead to the common lead and large focus is also measured in reference to the common lead. Normal resistances are very low (i.e., approximately 1 ohm or less), small focus being less resistive than large focus. A high resistance reading or an "open" would indicate a defective filament.

If the tube checks out mechanically, the service engineer should next connect the high voltage cables to the tube and then power-up the x-ray generator. With most generators, when power is turned on and no exposures are being attempted, the filaments are in a mode of operation termed ***standby mode.*** Depending on which focal spot size is selected at the console, a corresponding filament should "light" inside the x-ray tube. Both the small and large focus should be inspected for uniform brightness along the entire length of the filament wire. The filaments should then be "boosted" by hitting the prep switch and any defects, such as filament "sagging" or dark segments along the filament wire, should be noted.

This concludes the x-ray tube prestaging tests. The x-ray tube output and resolution are checked by taking actual exposures later in the final

13. The high voltage cables may have to be connected to satisfy the control logic of the x-ray generator. The service manual should be consulted before proceeding with this step.

14. A test connector can be made by using an old high voltage cable. The cable is cut 12 inches back from the federal connector end and the insulation of the three conductors should be stripped back. The cable end is inserted into the cathode end for testing and the continuity of the filament wires is measured with an ohmmeter between XS-XC and XL-XC.

stages of the installation. If, however, there is time (i.e., the room preparation is not yet completed), the engineer can perform a preliminary x-ray calibration (see the calibration section of this chapter). The collimator or a lead blocker must be placed on the x-ray tube port before proceeding with the calibration and the tube should be seasoned as outlined in the calibration section.

With the completion of the generator prestaging, all other equipment that will be installed in the radiographic room, such as x-ray tables and buckys, can now be tested for proper operation. Again, it is vital to know if any parts are defective before they are transported to the installation site.

EQUIPMENT DELIVERY

When the room preparation is completed and all the prestaging has been performed, the equipment is ready to be delivered to the site. The smaller items should be placed back in the original packaging for safe transport. It is not practical to recrate the larger items because of space limitations; however, all equipment should be wrapped with padding to prevent dents or scratches to surfaces.

The x-ray service engineer should be present at the site when the equipment arrives. This is important because the inventory should be verified and all items checked for damage. Also, the engineer can direct the movers, if they are not x-ray service engineers, in the specific placement of the individual devices. Heavy equipment such as the high voltage transformer should remain on dollies for the time being.

If an ***overhead tubecrane*** is being installed, it is usually mounted at this time with the help of the moving crew, the reason being that the heavy lifting equipment (e.g., a highjack) is now at the site and there is more manpower present to help with the lifting. The overhead long rails should have been cut to size and mounted to the ceiling supports by this time.

If there is enough space at one end of the rails, the complete tubecrane assembly, with the cross bridge assembly attached, can be lifted into position and then slid directly into the ceiling rails. Because of the size limitations of many rooms, however, this ideal situation for mounting the bridge does not always exists. Usually the cross bridge will have to be detached from the tubecrane and placed directly into ceiling rail.

The tubecrane assembly is then reattached to the cross bridge assembly after the bearings are properly adjusted.

The installer must check that the ***bearings*** are adjusted correctly at this time so that the tubecrane travels smoothly along the entire length of the rails. This is a very important step because the bearing adjustments are not always accessible when the tubecrane is positioned in the rails. Also, it is very difficult to take down (or "drop") the tubecrane when the rest of the equipment is fully installed in the room.

If it is decided that the tubecrane does have too much play (side to side or in the up/down direction), it can then be easily dropped and adjusted at this time while the highjack is still on the site. The leveling of the long rails should be checked at this time to ensure that the tubecrane does not drift along the length of the rails. If so, the rails attached to the ceiling must be "shimmed" until the tubecrane remains at rest when positioned at any point along the rail.

If a ***tubestand*** is being installed, it should be taken into the room by the movers and temporarily mounted at its specified location. This is done so that the correct cable length can be pulled to the tubestand. Since several accurate measurements must be made before the tube support can be permanently fastened in place, the floor and ceiling tracks should only be fastened with the minimum amount of hardware to hold it safely in place. During the alignment procedure, the floor and ceiling track will be repositioned and then permanently fastened. Important to note is that the location of the tubestand will dictate where the table and wall receptor will be located (see below).

Next, the x-ray generator control cabinet and the high voltage transformer should be positioned at the location specified in the blueprint. These units are usually positioned against a wall and will not be in the way of the installers. The transformer can be left on a dolly so that it can be moved when making cable connections. If a separate control console (or operator console) is used, it is usually located in the operator's booth and is mounted on a shelf, or on a pedestal stand if one was ordered with the system

The x-ray table should be placed as close as possible to its specified location for several reasons. First of all, because of their excessive weight, many x-ray tables are difficult to reposition without the proper moving equipment. In addition, if the table has to be moved, the flooring usually will get marked up with deep scratches that must be covered up cosmetically. Another reason for placing the table in its approximate location is so that the proper length of cables for table power and bucky control can be determined. This is important because all the cables will be "pulled" to each device in the next phase of the installation.

If the room layout was planned correctly, the table will be positioned in such a way that the tabletop can easily be removed without having to move the table base. The table top must be removed to properly install, align, test, and later service the internal components of the table.

When all the equipment is correctly positioned in the room, the service engineer can get a good "feel" as to the final appearance of the completed room. This is important because if some item was omitted or not correctly listed on the blueprint (i.e., a heating vent or radiator that was not listed, or a door opening outward versus inward), the room may not operate as promised. If last minute changes need to be made, the installer can make the appropriate changes to the actual room layout before each device is permanently fastened in place.

EQUIPMENT ALIGNMENT AND MOUNTING

With the equipment delivered and the moving crew departed, the actual physical installing of the x-ray equipment can begin. The main goal now is to get every device in the room aligned properly and then permanently mounted in place. In order to accomplish this, the x-ray tube and collimator must be installed on the tube support since they will be used to align the image receptors. Once the tube and collimator are mounted, the installer (a.k.a. the assembler) must get power to each unit of the x-ray system as quickly as possible so that the alignment procedure can begin.

Mounting the X-ray Tube

There are two types of x-ray tube housings that are used in general radiography. One type of housing has a raised mounting surface with four predrilled and tapped (1/4–20) bolt holes. The raised surface attaches directly to the tube support by threading the four Allen head (or hex head) bolts through the mounting plate of the tube support directly into the four threaded screw holes provided on the x-ray tube housing.

The second type of x-ray tube housing is mounted to the tube support by the use of trunion rings. The trunion rings of the tube support attach to the two raised (circular) ring surfaces on the tube housing. The trunion rings allow for tube rotation for special radiographic views. With the trunion mount, the tube is placed in the bottom half of the trunion ring assembly and then the top trunions are positioned on the top of the x-ray tube and then secured with four Allen bolts.

Installing the Collimator

The collimator, as stated in Chapter IV, is the beam limiting device for the x-ray tube. It is mounted either directly to the x-ray tube or to the tube mounting plate of the tube supporting device (Figure 22). Most collimators use a swivel type mounting bracket so that the collimator can be rotated. The most important part of mounting the collimator is determining the number of spacers that must be placed between the collimator and the x-ray tube!

The proper spacing is needed to provide the correct focal spot to collimator distance as specified by the manufacturer of the collimator. If the incorrect number of spacers is used, the light field dimensions will not correlate to the x-ray field dimensions. Light field to x-ray field congruence is a requirement of all collimators and will be verified during the collimator set-up procedure.

Also, if SID indication is provided on the collimator, it will not be accurate unless the correct number of spacers is used. The information packet provided with the x-ray tube contains a measurement of the location of the focal spot in respect to the port of the x-ray tube (where the collimator will attach). This dimension is used to determine the correct number of spacers required.

The spacers are provided by the manufacturer of the collimator and are inserted in between the collimator upper swivel bracket and the x-ray tube port. ***These spacers must be made of steel!*** If an alternative spacer, such as aluminum, is used, lead must be used to line these spacers. If the tube is mounted directly to the mounting plate on the tubestand, the thickness of the plate must be considered when calculating the spacer dimension. Once the spacers and upper swivel bracket are in place, the collimator itself can be attached to the upper bracket.

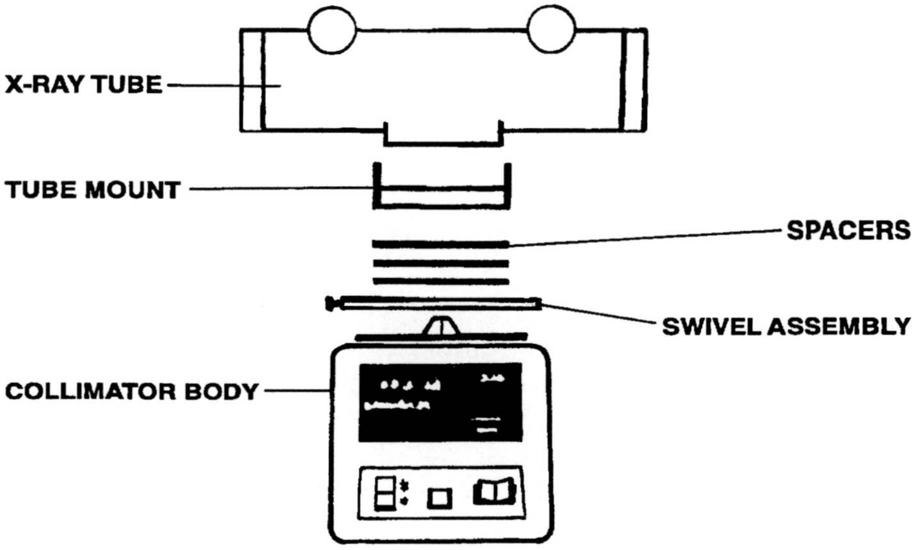

Figure 22. Collimator mounting. The correct number of spacers are required to achieve accurate SID indication.

Supplying Power to Each Unit in the System

To get power to the individual units, the service engineer must *pull and terminate all of the equipment cables*. Since each unit is in its approximate location, the amount of cable needed to be pulled can easily be determined. Once the cables are terminated and power is applied to the tube and collimator, the installer can proceed with the alignment procedures.

Types of Equipment Cables

There are basically three types of cables used in x-ray equipment: ***power cables, signal cables,*** and ***special high voltage cables.*** Each electrically operated device in an x-ray room requires at least one power cable, but most devices need both power and signal cables.

Power cables that connect the x-ray generator to the incoming power through a circuit breaker are made of heavy gauge wire (6 AWG, for example) surrounded by a thick insulating jacket. These power cables usually are not terminated (i.e., they have bare wire at each end) and are directly fastened to a terminal block. The cable used to connect primary power to the high voltage transformer is also made of this heavy gauge wire.

The service engineer must make sure the wire is the ***correct gauge and type of wire*** so that the equipment will operate safely and at the manufactures specifications. An underrated power line will greatly reduce the output of the x-ray generator because of excessive ***line loading.*** The specifications for wire size and type are found in the service manual for the equipment.

Each electronic device in the room will also have a power cable. These power cables, however, supply 120 VAC or low voltage DC power to the individual devices in an x-ray room and are usually made of much smaller gauge wire (e.g., 10–12 AWG wire).

Signal cables are used to interconnect or ***interface*** specific x-ray devices in the room so that they can communicate with each other and thus perform as a system. These cables are usually made of much smaller gauge wire (12 AWG or smaller) surrounded by a conductive cover or braid for shielding purposes, and enclosed in a flexible outer jacket. Signal cables are usually terminated with appropriate connectors at the factory and are simply plugged directly into each specific device during the installation. Care must be taken when pulling these cables through the conduit so that the connectors are not damaged.

The ***high voltage cables*** are a special type of power cable that connects the high voltage transformer to the x-ray tube. These (two) cables are designed to carry extremely high voltages and are, therefore, heavily insulated. Besides being thicker than most cables, high voltage cables are terminated with a special plug called a ***federal connector*** (Figure 13). Because of the size of the federal connector, high voltage cables are the most difficult to pull through conduit.

Pulling Cables

All cables are run inside of special conduit, called cable ***trough*** or ***raceway,*** which should have been installed during the room preparation phase. These cable "passage ways" are located along walls, above ceilings, or under floors. The task of pulling cables is much easier if the troughing has removable covers. In this case, all the cables are simply placed in the trough and the covers are then fastened in place.

Nevertheless, pulling the cables through conduit in the ceilings and floors is very difficult and usually requires two service persons. An ***elec-***

trician's fish tape or "wire fish," is used to pull the cables through the conduit and ***cable grease,*** available at electrical supply stores, is used to help the cables slide more easily through the raceway. If the cables from the previous installation are still present inside of the conduit, they can be used to pull the new cables through. This is not always possible because the older cables are usually removed during the room preparation phase.

Procedure for Pulling Cables with a Fish

1. Insert the wire fish at the destination end of the cable run (where the cable is to go) and feed it to the point where the cables originate (e.g., the power source).
2. Attach the cables *securely* to the fish with electrical tape. Make sure the cable is properly secured so it will not become detached while pulling. If the cable is terminated with connectors, these should be carefully taped so they are as streamlined as possible.
3. Apply cable grease at the point where the wire fish meets the cable. If other cables are in the raceway, apply grease to the entire cable to make pulling easier.
4. Gently pull ("fish") the cable through with the fish tape while the second person pushes or "feeds" the cable from other end.

This procedure should be followed for all the cables in the room. As a rule, power cables should be separated from signal cables in the trough to eliminate future problems caused by line "noise." Usually dividers are present in the trough to shield the individual cables. Once the task of cable pulling is completed, the cables must be properly draped before they can be terminated.

Cable Draping

Cables that are run externally to the ceiling mounted tubecrane, or to the floor to ceiling mounted tubestand, must be ***draped*** so that the tube support can move freely while having no stress applied to the cables themselves. If the cables are not draped properly, the tube could have limited motion in certain directions. More importantly, improperly draped cables will receive undue stress as the device is moved through its range of positions. A cable that is constantly stressed or overflexed will prematurely break and will have to be replaced. The goal for the service engineer is to drape the cables so they move as little as possible.

The cable drape must also be aesthetically pleasing to the eye since they are in plain view to anyone in the room. An x-ray room with poorly draped cables leaves an overall bad impression of the room. If the cables run along a ceiling rail, they should be draped so that they don't stand out and move fluidly with the device to which they are attached. If the cable run is very long, a method of draping the cables using ***coils*** or ***loops*** is often employed. If coils are used, attention should be paid to the size and number of coils, again for aesthetic reasons.

The process of draping cables properly is an art that takes time to develop. It often takes several attempts at draping just one cable to find the best position for that cable. As a result, cable draping is very time consuming and the service engineer must possess great patience to complete this task successfully. It is a process that cannot be rushed!

Cable Termination

Once draped, the cables should be terminated and connected to the appropriate equipment. The service engineer "terminates" a cable by adding the appropriate connector (or terminal) to the ends of the wires. These connectors are most often fastened to the wires by using a crimp connection. As stated earlier, some cables are terminated at the factory and, therefore, are simply plugged into their designated connectors.

When connecting power cables, strict attention should be paid to polarity and grounding. In addition, a phase rotation meter (or oscilloscope) must be used to achieve the correct line phasing when installing three-phase generators. Also, the x-ray generator can be used to switch power for the auxiliary equipment in the room, if desired. The collimator, tube support, and table power can be wired directly to terminals in the generator, if provided, so that the equipment in the room can be powered-up by just switching the main on/off switch at the generator control.

Every cable should be connected at this point in the installation. The high voltage cable ends should be coated with vapor proofing (see Chapter VII) compound before they are connected to the x-ray tube. Each cable should be identified, labeled with an adhesive wire marker if not already done, and then checked to see that the wire numbers are in agreement with the schematic. Many times the wire markers will be removed accidentally when the cables are pulled through the conduit. If this happens, the wire should be relabeled.

The installer should make an equipment interconnect diagram showing any special interfacing that was performed and include it with the service manual. The service engineer who will service this system in the future will appreciate the extra step taken by the installer.

Counterbalancing

When the x-ray tube and collimator are mounted, and the cables are connected and properly draped, the tubestand or tubecrane can be counterbalanced. As stated in Chapter V, the tube support must be counterbalanced so that the tube can be positioned in the vertical direction with minimal effort. The most common counterbalancing system consists of steel cables, pulleys, and special counterweights (usually made of lead). This type of counterbalancing system is used on tubestands and wallstands.

By adding counterweights supplied by the manufacturer, the tube can be balanced so that it will not drift up or down when the vertical locks are released. The tube should not drift at any point over the entire range of travel. Ideally, the tube should move up and down with very little effort and should come to rest immediately when released.

A different method of counterbalancing the x-ray tube is used in overhead tubecranes. Instead of being counterbalanced with weights, a spring mechanism or ***counterpoise*** is used. This mechanism is adjustable within a limited range. In addition, a compartment is provided in the tubecrane assembly so that the installer can add additional weights if the spring tension cannot be reduced enough for the particular x-ray tube.

Light Field/X-ray Field Verification

Before permanently fastening each device in the radiographic room, the engineer must first precisely align the x-ray tube to the image receptor(s). After all, a primary goal of the radiographic installation is to have the x-ray beam shooting directly at the film. Once the tube/receptor alignment is achieved, each device in the room can then be permanently mounted in place. Essentially, the radiographic room is installed around the tube/receptor axis.

However, in order to align the x-ray tube to the receptor, the installer must know the exact direction of the x-ray beam, or ***central ray.*** Since x-rays are invisible, the collimator light field is commonly used for this alignment procedure. The installer must, however, first verify that the light field of the collimator correctly matches the actual x-ray field so that he or she can obtain an accurate tube/receptor alignment. This means that x-ray system must be powered-up so that exposures can be taken.

Several exposures will be taken to verify the light field accuracy. In most cases, the film processor is not operational this early in the installation. Therefore, a ***fluorescent test screen*** should be used to visualize the x-ray field. The intensifying screens used in a 14 × 17 inch film cassette can be used if no test screen is available.

Since the x-ray tube has not been calibrated yet, a low technique must be used to prevent any damage to the tube. As a rule, the smallest mA station for large focus should be selected, using 60 kVp and an exposure time that will allow visualization of the x-ray field (at least 1/2 sec).[15] With the lights dimmed, an exposure is taken while viewing the fluorescent screen. (Note: the generator should have been configured for the line and type of x-ray tube during the prestaging phase. If generator prestaging was not performed, it must be done before exposures are taken!)

The collimator light field must accurately match the actual x-ray field within acceptable limits (i.e., 2% of the SID being used). To check the light/x-ray field, turn on the collimator lamp and close the shutters down to a square approximately 10 × 10 inches with the tube 40″ above the test screen. Clearly mark the four edges of the light field with metal markers.

If there is a noticeable discrepancy (i.e., 0.80 inches total @ 40 inch SID) between the light field (metal markers) and x-ray field, the engineer should make the required adjustments described in the collimator alignment section later in this chapter.

When the collimator light field check is completed, the equipment alignment procedure can begin. Before performing the alignment pro-

15. Some manufactures recommend derating the 200L mA station when making test exposures on an uncalibrated x-ray tube. This is accomplished by reducing the filament voltage for the mA station. The 200L station will, in actuality, produce much less then 200 mA.

cedure, the engineer should adjust each bearing in the tube support for smooth and level movements in all directions. *Any bearing adjustments must be checked at this time because they will affect tube alignment.* The tube supporting device must first be aligned to the wall receptor (if applicable). Once this alignment is achieved, the table is centered to match the tube/wall receptor axis.

This is an important phase in the installation because after this stage is completed, the major physical part of the installation is over, and the engineer can concentrate on calibrations and adjustments. Also, since a lot of dust and noise is created while the installer is drilling holes and mounting equipment, it is best to complete the physical mounting of the equipment as soon as possible. The staff at the facility will certainly appreciate the extra effort to complete this phase. After the equipment is mounted and the room thoroughly cleaned, it will begin to look like a completed x-ray room.

Overhead Tubecrane/Wall Receptor Alignment

The overhead tubecrane travels on ceiling rails that run along the length of the x-ray room. These ceiling rails should have been installed by the contractor so that they run perfectly parallel with the length of the room, as specified on the blueprint. Since the tubecrane is mounted directly to these rails, it should also run parallel with the length of the room. If the rails are misaligned, the tubecrane will not align to the wall receptor. This geometry is crucial since the tubecrane has very little adjustment for alignment in this plane.

A way to check the tubecrane alignment is to angulate the x-ray tube so that the x-ray beam is directed towards the wall where the receptor will be located and then to lock it in place. All other locks are then activated except the longitudinal locks. The tube is moved as closely as possible to the wall and the collimator light is turned on.[16] A mark is made on the wall at the center of the collimator crosshair pattern that is projected onto the wall. The tubecrane is then run along the railings, away from the wall, while the service engineer observes the light field

16. Bearing adjustment should have been verified before proceeding with alignment check. Also, the tube should be locked in the detent for rotational movement.

in relation to the mark on the wall. The crosshairs should remain on the mark and should track accurately for the length of the rails.

If the tube does not track in the left to right direction, then the ceiling supports were not installed correctly and the long axis of the equipment will be slightly different from the axis of the actual room. If the alignment is off by a small degree, it may be possible to compensate for this error by skewing and shimming the wall receptor. If the rotational detent of the tubecrane can be adjusted, the tube assembly should be rotated in the direction of the error. Aesthetically, however, the room will look awkward and unprofessional.

If the tube does not track in the vertical direction while moving along the rail, this indicates that the ceiling is sloping. Minor corrections can be made by shimming the ceiling rails. Sometimes it is impossible to achieve perfect tracking in the vertical direction for the entire length of the room. In this case, the engineer should then concentrate on aligning the tubecrane in the usable range for the wallstand (40″–72″ from wallstand). Once the tubecrane is perfectly aligned to the wall, the installer can permanently mount the wall receptor (see "Mounting the Wall Receptor").

Floor to Ceiling Tubestand/Wall Receptor Alignment

Aligning a floor to ceiling tubestand to a wall receptor is more involved than the tubecrane alignment procedure. The reason for this is that the tubestand must be aligned in three planes: the longitudinal plane, the vertical plane, and the horizontal plane (Figure 15). To accomplish this alignment, the procedure described below must be followed in the exact order given.

The first goal is to align the tubestand ceiling track so that it runs perpendicular to the plane of the wall receptor. If the room is perfectly square, the walls can be used as the reference for measurements. Another way to find the position for the ceiling rail is to use a large carpenter's square and mark off a right angle at the receptor wall and then extend the marks with a chalk line for the entire length of the track.

With some tubestands, the upper guide rail can be mounted directly to the back wall, eliminating one of the variables (i.e., the longitudinal plane). In this case, the wall-mounted track is fastened into place and then a plumb bob is used to mark the position of the floor track.

To align a tubestand:

1. Mount and secure the ceiling (or wall) track first, following the procedure described above.
2. Mark the floor using a plumb bob referenced to the upper track and lay floor track on the floor, but do not fasten.
3. Level the tubestand column vertically along entire length of track using a four foot level. Check for level in both the front to back and side to side directions.
4. Once vertical column is level, check the level of the horizontal arm which holds the x-ray tube, and adjust the bearings so arm and x-ray tube are level.
5. Angle the tube to the wall where the wallstand will be located. Turn the collimator lamp on and make a mark on the wall at the center of the projected crosshair. (Tape a piece of paper to the wall to protect the wall.)
6. Run the tubestand along the entire length of the track. Notice if the light field drifts (or "walks") off of the mark. If the projected crosshair stays on mark, then the alignment is completed, and the floor track can be permanently fastened to floor.
7. If the light field drifts from the mark, recheck the alignment of the upper and lower track. If correct, then the last adjustment to be made is to the tubestand rotational detent.

The tube rotational adjustment is always made in the direction of the error. Make small adjustments to the detent and check with the light field to see if alignment improves. A new mark must be made on the wall each time the detent is changed. When the light field is tracking perfectly on the wall mark, the tubestand alignment is complete.

Mounting the Wall Receptor

If a wall receptor is being installed in the general radiography room, it should be permanently mounted at this time. The preferred setup is to have the wall receptor mounted at the head end or foot end of the table, aligned with the center axis of the table. With this set-up, once the tube is locked at table center, the tube will be centered automatically to the wall receptor by simply angulating the x-ray tube toward the receptor.

Since the tube was aligned to the wall earlier in the installation, mounting the wall receptor is a simple process (Note: If the tube support has cross-table/lateral motion capabilities, position the arm at the center of travel.):

1. Angulate the tube toward the wall where the receptor will be located.
2. Turn on the collimator light.
3. Position the center line of the wall receptor at the center of light field.
4. Verify tracking from the wall unit out to 72″ SID. Shim, if necessary, and mount to floor and/or wall.

Wall Receptor Adjustments

The wall receptor should be checked for smooth vertical travel. If a wall stand is being installed, the engineer should pay particular attention to the ***counterweight assembly.*** The correct number of lead counterweights must be installed or smooth and safe vertical movement. When adding weights for counterbalancing, the installer should make sure that the cassette tray, loaded with a 14″ × 17″ film cassette, is inserted into the bucky. The 14″ × 17″ film cassette is used because it is the most common type used for erect studies.

The vertical lock assembly (friction lock or electromagnetic) should be inspected at this time and adjusted, if necessary. If a wall bucky is used, the reciprocating grid should be tested following the procedure for the table bucky described later in this chapter.

Aligning and Mounting the Table

Once the tube support has been permanently mounted, the table can be positioned and then fastened to the floor. The tabletop should be removed for this procedure since the point of the table alignment is to match the x-ray field to the table receptor (i.e., to the film). Although some table manufacturers provide a centering detent for table top centering, this should not be used until its accuracy has been verified. By removing the tabletop, the light field can shine directly on the receptor for an exact alignment.

If a tubestand with cross-table capabilities is being used, the service engineer should make sure the lateral arm of the tubestand is at the center of its travel. With the arm in the center detent, the x-ray tube should be centered with the wall receptor (see above). Again, the center of the table receptor should perfectly align with the center line of the wall receptor.

With the tabletop removed and the receptor approximately in the center of its travel, proceed as follows *(Note: the table is being move to match the tube, which is already aligned to the wall receptor and permanently mounted!):*

1. Position the tube at 40" from the tabletop and center it to the receptor. Make sure the collimator is locked in its rotational detent. Energize all of the locks on the tube support except the longitudinal lock. Turn the collimator lamp "on."
2. Check the receptor at this time to see that it is not rotated in respect to the light field. If rotated, the installer should adjust the bearings on the receptor assembly.
3. Move tube and receptor together toward foot end of table. Slightly "bump" the table at the foot end until the light field and receptor match. Next move the tube and receptor to the head end of the table. If the light field does not track with receptor, "nudge" the table gently to achieve alignment
4. Continue to repeat step #2 until perfect tracking is achieved.
5. Mark the table location and permanently fasten.

Table Adjustment

While the tabletop is removed, the service engineer should perform the table adjustments. In fact, most table adjustments can be performed more accurately and with greater ease with the top off. Items to be adjusted include the receptor assembly, the electromechanical locks, the table travel limit switches, and the motor drive assemblies.

If a reciprocating bucky is being installed, the tabletop must be removed (or at least moved to one end of the table) so that the entire bucky assembly can be visualized. With the bucky fully exposed, the service engineer can easily inspect the motor and all connections of the bucky assembly. In addition, the engineer can energize the motor drive circuit and directly observe the actual movement of the reciprocating grid. If any problems are found, they will be much easier to repair at this time.

The operation of the reciprocating grid should be inspected visually and audibly. The movement of the grid should be relatively quiet and smooth. If any vibrations are heard, the engineer should check that all the hardware on the grid assembly is fastened properly. The grid must also move equal distances on either side of the bucky center line. To check this, the engineer should position the x-ray tube directly at the bucky center line as indicated with the collimator light field. Usually there is a mark on the bucky frame to indicate the "bucky center" position.

With the collimator field lamp on, the bucky circuit should be energized again. The center of the grid should travel an equal distance on either side of the center of the light field. If it doesn't, the switches responsible for bucky travel direction should be adjusted until equal travel in both directions is obtained.

Finally, the engineer should check and tighten all of the hardware within the bucky assembly. He or she should inspect the cassette tray for smooth travel in and out of the bucky. If automatic collimation is used, the cassette tray plug alignment should be checked. The plug on the tray should easily mate with the receptacle located at the rear of the bucky frame. The pin that activates the size sensing arm in the cassette tray should also be checked and tightened.

The tabletop electromagnetic locks (and wiring) should be visually inspected before replacing the top. The locks will have some up/down movement (spring action) but should not rotate more than a few degrees. Once the engineer has completed the visual inspection, the tabletop should be replaced and the lock adjustments checked.

The tabletop should move easily in all directions with the locks released. A grinding or scraping sound could indicate a lock that is dragging or misaligned. If a lock is, in fact, dragging, the distance between the lock and the contact surface (i.e., the metal strip on the underside of the tabletop) should be increased. The engineer should also ensure that all of the electromagnetic locks are activating and should adjust any that are not. This is usually done by decreasing the distance between the lock and the metal strip on the table top. A lock that is not adjusted properly can easily be detected by the loud humming sound that it makes. Any locks that fail to operate and cannot be adjusted should be replaced at this time.

The tabletop bearings should be checked and adjusted for the smoothest movement. If the tabletop is motor-driven, it should also be inspected and adjusted at this time.

All the mechanical parts of the table vertical drive system (i.e., gears, chains, and belts) should be checked and adjusted, if necessary. Also, all bearing adjustments should be checked at this time. The table should drive smoothly up and down with no apparent strain on the motor. The drive system for tilting tables, including all gears, belts, and chains, should also be inspected at this time and adjusted for smooth operation.

The limit switches for up/down and tilting motion should activate at the correct distances as specified by the manufacturer. Some tables provide an exposure position at a point midway between the full up and down position. This exposure position should be adjusted in conjunction with automatic collimation. Any safety interlocks for the table drive should be checked for proper operation.

Tube Centering Detents

Most radiologists and x-ray technicians request that centering detents be installed so that they will not have to visually align the x-ray tube to the receptor for each patient. A detent is a mechanical device used to stop the motion of the tube assembly. Spring-loaded, centering switches and detents are used to automatically center the tube over the receptor. The installation of the switch and detent for table centering is most easily performed with the tabletop removed.

While the tabletop is off, the installer can verify that the x-ray tube is correctly centered over the receptor. The engineer can then accurately mark and mount the switch and detent so that perfect tube/receptor alignment can be achieved.

If the table was aligned with the center of the wall receptor as described earlier, only the one centering switch is needed. If they are not in alignment, the same technique is used for mounting the centering hardware for the wall receptor.

CALIBRATION AND ADJUSTMENT

Installing a radiographic room is not a simple task. Installations are physically demanding and often involve working long hours and on weekends. In addition to the moving and lifting of heavy equipment that is required, the installer must pull, cut, crimp, "ring out," terminate, and label countless wires (sometimes in the hundreds). This is tedious work that is very time consuming. To make matters worse, the service engineer will often perform these duties while lying on the floor or while on a ladder up in the ceiling. Most of the time, the engineer must work in small, unlit areas–behind equipment cabinets, for example. This type of work can definitely take a heavy toll on the service engineer, and a change of pace is welcomed.

The next phase of the installation is the calibration and adjustment phase. The goal now is to get all the devices in the radiographic room "set-up" and calibrated to the manufacturer's standards. During this phase, the rotor controller is adjusted, the x-ray tube is calibrated, the collimator is adjusted, and the AEC is calibrated.

It is often the case that if a piece of equipment is going to fail, the failure will occur during the calibration and adjustment phase. When calibrating the x-ray tube, for example, many exposures must be taken often at near maximum tube and generator ratings, at techniques that would not normally be used for most x-ray exams. Moreover, the other equipment in the room is repeatedly maneuvered during the adjustment phase, simulating the "slamming car door" scenario seen in automobile testing facilities. When this stage is completed, the room will definitely be ready for use.

The first priority of this phase is to get the x-ray tube calibrated, since many exposures will be taken when adjusting the collimator and AEC circuits. Calibrating the x-ray tube now will ensure that the tube s being operated safely during those adjustments. Before starting the calibration, however, two important steps must be taken. First, the rotor controller has to be properly set up, and second, the x-ray tube must be ***conditioned*** (or ***seasoned***). If these steps are not taken, a stable calibration will not be achieved and, even worse, possible tube damage could occur.

Starter Verification

As stated in Chapter III, a rotating anode is basically a split phase induction motor, consisting of a ***stator*** which is a set of stationary electromagnets, and a ***rotor,*** which rotates inside of the stator. The rotor controller, or ***starter,*** is a device that controls the rotation of the anode in a rotating anode tube. It is basically a single phase AC power supply that supplies current to the three stator windings: stator main, stator phase-shift (start), and stator common.

As with all single-phase induction motors, the main winding and the start winding must be out of phase in order to initiate rotation. To achieve this phase relationship, a ***capacitor*** is placed in series with the start winding at the rotor controller. The degree to which the main and phase-shift windings will be out of phase is directly related to the size of the capacitor used.

A ***Low Speed Starter*** operates by applying a ***starting voltage*** (approximately 100–240 VAC) to both the main and phase-shift windings of the stator. The current in the phase-shift winding, which has a capacitor in series, leads that of the main winding (see Figure 11). The current moving through these windings, by induction, forces the anode to rotate and accelerate up to the correct speed (approx. 3,300 RPM). After the preset time delay, the anode will be rotating at the proper speed and the starter supply is then switched, through a set of relay contacts, to a ***running voltage.*** This running voltage, of approximately 40–80 VAC, is used to maintain rotation of the anode until the exposure is completed. At the end of the exposure, the tube coasts down slowly until it stops.[17]

Starter verification consists of first measuring the start and run voltages (preferably with an oscilloscope in order to observe the phase shift) and comparing the voltages with the manufacturer's recommendations. Once this is done, the rotational direction of the anode (CW or CCW as viewed from the cathode) must be verified. Finally, the actual rotational speed is measured with a ***vibrating reed tachometer.***

A vibrating reed tachometer accurately measures the rotational speed of the anode and should be used during the installation. As the name implies, it measures the speed by sensing the frequency of vibrations in the x-ray tube housing and then displaying the value in rpm's. The reed tach is place directly on the tube housing during the boost and run cycles and the actual speed of the anode is read by the engi-

17. A DC brake is used in high speed starters only. Its purpose is to quickly reduce the speed of the anode so that it will not rotate through the resonant frequencies of the x-ray tube. If the anode rotates at resonance, damage will result to the glass insert. Resonance is never reached with low speed operation.

neer. This measurement is very important because if the tube is not rotating at the correct speed, the heat rating of the tube will be reduced and target damage can occur.

The rotational speed can be adjusted by the service engineer. Adjustment is accomplished by increasing or decreasing the "boost time" (i.e., the time that starting voltage is applied to the stator windings). Another way to adjust anode speed is by increasing or decreasing the starting voltage.

Conditioning an X-ray Tube

All new x-ray tubes, and tubes that have not been energized for an extended period of time, must be ***conditioned*** (or ***seasoned***) before they can be calibrated and then used on actual patients. A tube that has been stored (i.e., not been energized) for as long as one month will develop an imbalance of internal charges along the inside surface of the glass insert. This charge imbalance will result in erratic exposures, making it impossible to achieve a stable calibration.

The purpose of "seasoning" a tube is to realign the distribution of electrical charges within the glass envelope so they are properly matched to the x-ray system. When the correct distribution has been established, the output of the x-ray tube should be stable and the service engineer can then proceed with the x-ray calibration. *The service engineer should never proceed with x-ray calibration of a new tube until the tube is seasoned!*

Many service engineers will start the x-ray calibration immediately, in an attempt to season the tube during the calibration. This practice generally creates more work for the engineer and should be avoided. If the service engineer attempts to calibrate a tube that has not been properly seasoned and a tube arc occurs, the calibration procedure must be stopped in order to perform the tube seasoning procedure, anyway. When the tube is completely conditioned, the engineer must start calibration over again.

Furthermore, *the x-ray tube must be at near room temperature before proceeding with the tube seasoning procedure.* If a replacement tube has been transported from outdoors during the winter, for example, and is at a temperature near freezing, exposures cannot be taken until the temperature of the x-ray tube stabilizes to near room temperature.

If the engineer is under time constraints to get the room operational, the tube can be "warmed-up" more quickly by connecting the high voltage cables and the rotor cable, and then powering-up the x-ray unit. With the generator power on, the equipment will be in a "standby mode," in which case the selected filament of the x-ray tube will have a small amount of current flowing through it and will be dimly lit. The heat given off from the filament will help warm up the tube. While in standby mode, the engineer should alternately select between large and small focus at the x-ray control so that both filaments will be heated.

If really pressed for time, the engineer can select a low technique on the console, press and hold the prep button for several seconds, and then release the prep switch for 30 seconds ***(DO NOT TAKE AN EXPOSURE!).*** By performing this procedure, the filaments are raised to a higher temperature which adds heat to the tube. In addition, the process of boosting the tube also energizes the rotating anode which helps introduce heat into the tube housing. The "prep and release" cycle should be repeated for several minutes while the engineer alternately selects the large and small focuses.

The engineer should periodically check the temperature by placing his or her hand directly on part of the housing. When the housing temperature approaches that of the room, exposures may be taken at very low techniques and spaced a minute apart. After these steps have been taken, the tube should now be ready for the seasoning procedure.

The actual process of tube seasoning involves taking successive exposures starting at a low kVp (low mA) and slowly raising the kVp to maximum value. It is always recommended to follow the x-ray tube manufacturer's guidelines for conditioning an x-ray tube; however, a general tube seasoning procedure that will safely work for radiographic tubes is provided below. (Note: the tube current should be monitored during this procedure to ensure that it is within safe operating limits-see calibration.)

1. Select the lowest mA station possible for large focus.
2. Select 60 kVp and an exposure time of 1/4 second (250 ms).
3. Take 3 exposures–30 seconds between exposures (observe MA waveform).
4. Repeat this process increasing by 10 kVp until maximum kVp is reached.

If arcing occurs at any point in this procedure, back up to the previous kVp setting and repeat the exposure sequence. When tube seasoning is completed, the x-ray tube calibration can be performed. If for any reason the x-ray tube cannot be calibrated directly following the tube seasoning procedure (e.g., if hospital personnel are in the room, lunch time, etc.), a tube warm-up is recommended before starting calibration.

Tube Warm-up Procedure

If the x-ray tube has been idle for 4 hours or more, it should be "warmed-up" before it is calibrated (or used on patients). During the tube warm-up procedure, exposures are taken using a low technique that will slowly apply heat to (i.e., warm-up) the entire surface of the target. Once the target is sufficiently heated, it can be safely operated using higher techniques. Also, by warming up the tube before calibrating, a more stable calibration will be achieved. As a rule, *a daily tube warm-up procedure will help promote longer anode life!*

If a cold target is "hit" with a high technique exposure, the resulting high tube current will quickly heat the area of impact on the target to a very high temperature. As a result, a large contrast in temperature now exists between the impact area and the rest of the target surface which is cold. The extreme variations in temperature across the surface of the target causes a condition known as "thermal stress."

Thermal stress will cause cracks or pits to form on the target surface, and may cause the anode to become warped. Any of these conditions will effectively make the tube useless. If the tube is used in this condition, erratic exposures will result and more damage could possibly occur to the x-ray generator. Consequently, a tube with target damage will have to be replaced.

The warm-up procedure is simply a shorter version of the tube seasoning procedure and is listed below:

1. Select the lowest mA station for large focus and 1/10 second exposure time.
2. Take one exposure every 30 seconds at: 80 kVp, 90 kVp, 100 kVp, and 120 kVp for a total of four exposures.

This procedure is sufficient to warm up most diagnostic x-ray tubes. When used daily by the x-ray technicians, this will absolutely increase the life of the x-ray tube.

Calibration Measurement Techniques

Once the x-ray tube conditioning has been completed, the service engineer can proceed with the x-ray tube calibration. An accurate calibration is required to ensure safe operation of the x-ray tube and also to meet BRH standards (Bureau of Radiation Health) which are specified by law. Moreover, a properly calibrated x-ray system will deliver consistent exposures, which are required to produce good diagnostic quality images.

In order to calibrate an x-ray system, the service engineer must obtain accurate measurements of the actual x-ray tube voltage (kVp), tube current (mA), and exposure time for a given exposure setting. While making these measurements, adjustments are made so that the nominal values that are set on the x-ray control will agree with the actual values read on the test equipment. The measurements of kVp, mA, and time can be obtained in two distinctly different ways: ***invasively*** or ***noninvasively.***

Invasive vs. Noninvasive Measurements

Invasive measurements are obtained by inserting the test equipment "in circuit" to measure the actual tube current (mA) passing through the tube, and the actual peak kilovoltage (kVp) that is applied to the x-ray tube. With invasive measurements, the test equipment is wired in series with the x-ray tube. Consequently, the values obtained by invasive measurements are the most accurate since they represent the exact voltage and current that the tube is actually "seeing." *A calibration made with invasive measurements will be the most accurate.*

Noninvasive measurements, on the other hand, are made by placing a detector in the radiation field. The detector senses and filters the radiation output from the x-ray tube and then calculates the x-ray tube voltage, mr output, and time from the resulting radiation waveform. What is being measured is the x-ray tube *output* and not the actual voltage *applied* to the tube. Since two completely different parameters are being measured, invasive measurements usually do not exactly agree with measurements taken noninvasively.

One reason for the dissimilarity between the two ways of measuring kilovoltage has to do with the fact that noninvasive measurements are greatly affected by many external factors. Those factors that most influence noninvasive measurements are ***detector positioning and the quality of the x-ray beam.*** Moving the detector to slightly different locations within the radiation field will greatly affect the final measurement values. In addition, if the detector is not placed precisely in the same position each time a calibration measurement is checked, the service engineer cannot rely on the results that he or she obtains. At best, if the positioning is held constant, the noninvasive measurements can be used as a reference to track calibration stability over a period of time.

The quality of the x-ray beam can also affect the readings obtained with a noninvasive meter. Specifically, the beam quality of the x-ray tube (see Chapter VIII) can affect the performance of the meter, itself. For example, the accuracy of a noninvasive meter is questionable when testing an older x-ray tube that has a significant amount of filtration. To make matters even worse, the kVp accuracy tolerance of most noninvasive meters is usually specified at 5 percent. This means that at a setting of 80 kVp on the x-ray control, the meter can read anywhere from 76 kVp to 84 kVp.

For the above reasons, the service engineer must always make invasive measurements during a calibration, especially for the initial calibration at the time of installation. Also, the engineer should never use a noninvasive kVp meter for diagnosing x-ray problems. The waveforms produced by the noninvasive meter only represent the radiation output level and *not* what is actually occurring in the high voltage secondary circuit.

A problem that service engineers face today relates to the fact that the radiation physicist who inspects the equipment uses only noninvasive equipment for kVp output measurements. If the engineer performs a calibration with invasive test equipment, he or she may get results that differ slightly from the physicist's results. Consequently, service engineers will tend to perform calibrations with noninvasive equipment so that their results are in close agreement with the physicists'. For the record, this is not a good practice.

As stated, the most accurate calibration will be achieved with invasive test equipment. In addition, by viewing the actual kVp and mA waveforms on an oscilloscope during the calibration, any abnormalities in the waveforms can be identified and noted. This is important because any problems that are identified now can be repaired while the equipment is new and still under warranty.

It is of the utmost importance for the engineer to know that the equipment is operating at its optimum level and that all the x-ray waveforms are correct at the time of installation. By knowing this he or she will have a reliable reference that will be used later when troubleshooting x-ray problems. Ideally, the kVp and mA waveforms should be documented and stored in the equipment file for that particular room.

There are two invasive test setups that are commonly used during an x-ray calibration. The first setup utilizes a ***Dynalyzer with a digital display.*** The Dynalyzer is a special (and most expensive type of high voltage divider that provides kVp measurements, as well as, mA, time, and filament current measurements. The Dynalyzer is installed in series with the x-ray tube (preferably at the x-ray tube) using the high voltage cables that are provided with the Dynalyzer.

The service engineer can read the measurements directly from the digital display and, if the Dynalyzer is used in conjunction with a digital storage oscilloscope (DSO), the actual kVp and mA waveforms can also be viewed for diagnostic purposes. Because it is highly accurate and is easy to setup (only one piece of test equipment), the Dynalyzer has become the most commonly used test device for x-ray calibration and is now used as the industry standard for calibration by many equipment manufacturers.

The second referred test setup used for x-ray calibration consists of a ***High Voltage Divider*** (or ***Bleeder***), a ***DSO,*** and a ***Digital mA/mAs Meter.*** The high voltage divider is a test device that contains a voltage order network (series resistors) immersed in insulating oil. It is connected in series with the anode and cathode cables at the x-ray tube exactly like Dynalyzer. The difference with the bleeder is that it is used for measuring kVp only and cannot measure mA. Since the kilovoltage applied to the tube is measured in ***peak value*** (hence kVp), it can be read directly from the waveform that is displayed on the oscilloscope. As with the Dynalyzer, the actual kVp waveform is also used for diagnosing high voltage problems.

High voltage bleeders are made by several manufacturers and are much more reasonably priced than the Dynalyzer. The drawback, if it can be considered one, is that a separate piece of test equipment (the mA/mAs meter) is needed for the measurement of tube current.

For the measurement of tube current when using a bleeder, a ***digital mA/mAs meter*** should be used. The advantage of using a digital meter is that it automatically converts and displays the ***average DC mA value*** (more on this subject later). The mA/mAs meter is connected in series with the mA metering circuit of the x-ray generator, usually found at the midpoint of the high voltage secondary (refer to Figure 4).

The digital mA/mAs meter has two modes of operation. In the ***mA mode,*** the meter reads the tube current (mA) in "real time" during to exposure. The problem with using the meter in this mode of operation is that fairly long exposures of approximately one second are required in order to obtain a stable mA reading. If a complete calibration was performed using the mA mode only, the x-ray tube housing would eventually overheat. In addition, the tube exposure limits would be exceeded at the higher mA stations and exposures would be blocked. Consequently, this mode can only be used for the lower mA stations (i.e., 25 mA, 50 mA).

To avoid any possible damage to the x-ray tube, the second mode of operation, the ***mAs mode,*** must be used when calibrating the higher mA stations. In the mAs mode, the meter reads the product of mA and time (mA-sec or mAs) and then displays the result. The service engineer will set a time on the x-ray control that will make calculations simplified, 0.10 seconds, for example. In addition, the mA/mAs meter is designed to hold the most recent mAs reading on the display until the next exposure is taken. Most service engineers own an mA/mAs meter because it is easy to use and is very reliable.

Using a High Voltage Divider

The manufacturer's guidelines are included with the high voltage divider and should be followed when connecting it to the x-ray system. Both the anode and cathode cables (provided with the bleeder) are initially connected to the x-ray tube in a ***series connection*** (see Figure 23). Prior to the actual calibration, several exposures must be taken at different kVp settings to verify that the anode and cathode kVp waveforms are balanced (i.e., equal in amplitude but opposite polarity).

Once this is verified, the engineer should turn generator power off and remove the cathode cable from the bleeder, discharging the pins to ground first, then reconnecting it back into the x-ray tube. Calibration can now be performed accurately using just the anode waveform. Of course, with this test setup configuration, the kVp value obtained will now be one-half of the actual kVp setting.

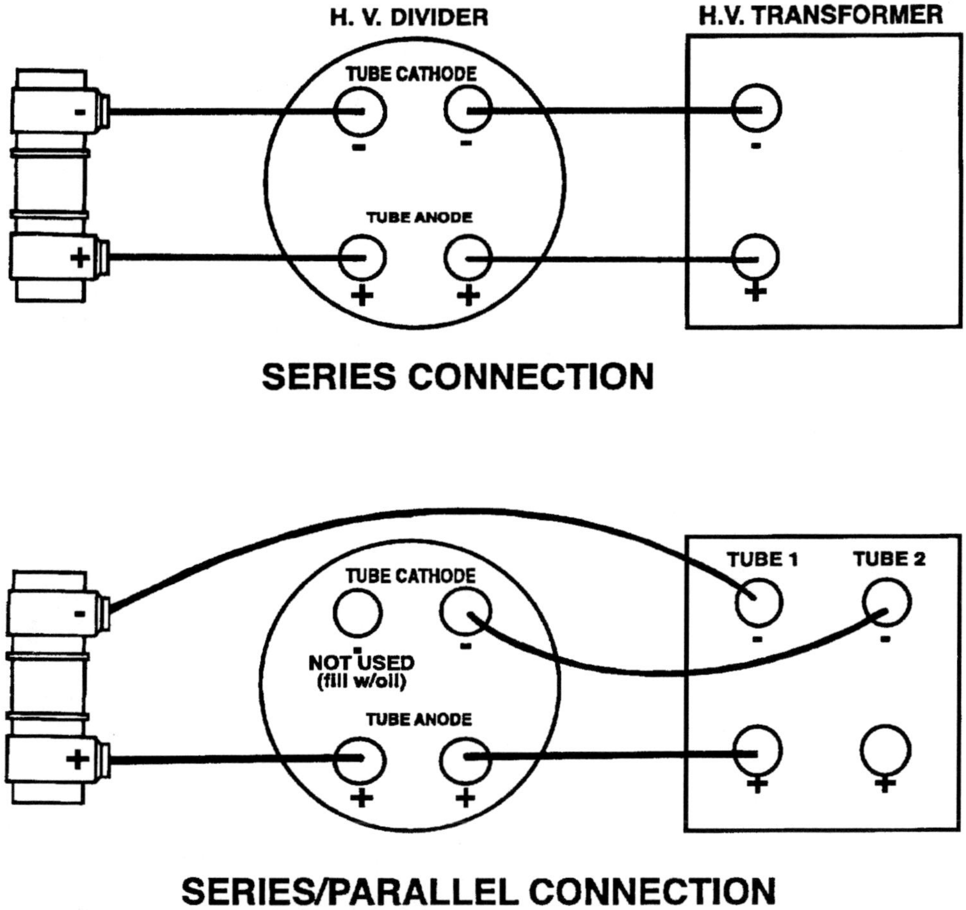

Figure 23. Connecting a high voltage divider. The series/parallel connection allows for viewing the cathode kVp waveform without introducing a load factor to the x-ray generator.

The reason for taking the cathode side connection of the bleeder out of the circuit has to do with a phenomenon called ***filament loading.*** The bleeder, with cathode connected, will add a small amount of series resistance to the filament circuit. If left connected during the calibration, the service engineer will adjust the filament voltages slightly higher than they would normally be set to achieve the desired mA reading. When the bleeder is removed after calibration, the series resistance is also removed and, consequently, the higher filament voltage will be dropped entirely across the x-ray tube. This will result in the actual mA value being slightly higher than the set value.

If the service engineer needs to view both the anode and cathode waveforms during the calibration, the mA's will have to be readjusted when the bleeder is removed. A correction factor, if known, can be used when calibrating with cathode connected. During the calibration, the engineer will automatically adjust each mA station a certain percentage lower.

This method of calibration is accurate if the exact correction value is known for that particular x-ray system and the specific bleeder being used. The only way to know this actual correction value is to calibrate the system with the cathode side removed (no filament loading) and compare those readings with the reading obtained with the cathode side connected.

If an additional set of high voltage receptacles is provided on the high voltage transformer, and is not being used, the high voltage test device can be set up in a ***series/parallel*** configuration by activating the second set of wells. This series/parallel set-up enables the service engineer to view both cathode and anode waveforms while calibrating without loading the filament circuit.

The Dynalyzer (or high voltage bleeder) should be used during the first calibration at the time of installation. Once the calibration is completed, and while the test equipment is connected to the system, the service engineer should perform two important checks. First he or she should place a ***noninvasive kVp meter,*** in the radiation field and compare the results with those obtained invasively. This will not only confirm the operation of the noninvasive device, but will allow the service engineer to actually see the degree of variation between the two methods of measurement. After performing this check, the noninvasive meter can now be used, with confidence, as a reference for calibration checks.

Secondly, the service engineer should confirm the accuracy of the mA panel meter located on the x-ray control. He or she should compare the readings obtained with the test meter to those of the panel meter. This is done so that if an mA calibration check is needed during a service call, the engineer can rely on the panel meter with confidence, thus saving valuable time.

If the x-ray generator has digital control circuitry, test points for kVp and mA waveforms are often provided. The calibration values obtained at these tests points should also be compared to the readings obtained with the test equipment at this time, and adjusted if necessary.

When the adjustments are complete, these test points can be used for calibration checks likewise eliminating the need to connect external test equipment.

The high voltage divider is such an important tool for servicing x-ray equipment that every engineer should have access to one. If, however, a high voltage divider is not available for the initial calibration, a noninvasive kVp meter can then be used. This test set-up should be used as the *last option only* for x-ray calibration. Also, the noninvasive kVp meter must be used in conjunction with a digital mAs meter for the most accurate calibration. The output radiation waveform should be monitored with an oscilloscope during the calibration. If any abnormalities in the waveform are observed, the engineer must use an invasive kVp divider in order to further diagnose the abnormal waveform.

If an mAs meter is out of the budget, there is another way to measure mA invasively. The service engineer can insert a resistor (usually 10 ohm, 10 watt) in series with the midpoint of the secondary of the high voltage transformer, at the point where the mAs meter normally would be placed. By connecting an oscilloscope probe across the resistor, a voltage waveform representing the actual mA can be obtained. Because this waveform represents the actual current flowing through the x-ray tube, it can also be used to check the exposure time and can be used for diagnosing x-ray problems.

There are, however, a few important considerations when using this test setup. The engineer must exercise great care when connecting the resistor and oscilloscope "in circuit" to ensure that they are isolated from ground.[18] Also, the engineer must be determine the type of high voltage transformer that is used in order to know what type of mA waveform (AC or DC) to expect (see below). Finally, with waveforms obtained with this setup, the actual mA numerical value cannot be read directly from the peak waveform on the oscilloscope. The peak reading must first be converted to its ***average value*** equivalent before making any mA adjustments. ***MA is measured in average value!***

The average value is an arithmetical average derived from one-half of a sine wave and is equal to 0.637. The average value is obtained by multiplying the peak voltage reading by the value of 0.637. This calcu-

---

18. Use caution when working in the mid-secondary of the high voltage transformer because of the dangerous potentials that exist there. Also, since the oscilloscope ground must be "floated," the chassis will become "hot." The newer hand held battery powered scopes are good for this application because they are isolated from ground.

lation can be applied to any AC (or pulsating DC) waveform to obtain the equivalent effective DC power rating, and must be used when calibrating single phase generators with the test resistor.

The average value does not apply to three-phase and high frequency generators because their output waveforms are essentially at a near constant DC level (with a small ripple component). With these units, the mA is read directly from the waveform on the oscilloscope by averaging out the ripple factor.

For single-phase generators, the waveform obtained across the 10 Ω resistor will be either AC or DC (with 100% ripple), depending on the configuration of the secondary winding of the high voltage transformer. For an AC waveform, the peak value (V_{peak}) of one-half of the AC cycle is used for further calculation. Or, if desired, a bridge rectifier can be installed at the midpoint to convert the waveform to DC. If the output waveform is DC, the peak value of the entire waveform can be used for calculation.

For example, if calibrating the 100 mA station, the engineer must read a one volt (average) drop across the resistor. Using Ohm's Law: 100 mA x 10 ohms = 1 volt (the value chosen for the resistor allows for convenient calculations). Since the engineer is viewing a peak voltage waveform on the oscilloscope, he or she must take the peak value and then convert it to its average value. To convert a peak reading to average, use the formula: $V_{average} = 0.637 \times V_{peak}$. For the example above, a value of 1.57 V_{peak} must be read on the oscilloscope to obtain 100 mA of tube current (1.57 $V_{peak} \times 0.637 = 1.0\ V_{av}$). A conversion chart for commonly used mA settings when using the 10 Ω resistor for calibration is provided below.

MA Volts Conversion Chart

25 mA	0.39 V_{peak}
50 mA	0.78 V_{peak}
100 mA	1.57 V_{peak}
150 mA	2.35 V_{peak}
200 mA	3.14 V_{peak}
300 mA	4.71 V_{peak}
400 mA	6.28 V_{peak}
500 mA	7.85 V_{peak}

It is important to note that the average value calculation applies, theoretically, to a perfect sine wave only! Technically speaking, the mA waveform obtained at the midpoint of the secondary is slightly distorted due to the capacitance of the high voltage cables and will, therefore, not be perfectly sinusoidal in shape. Consequently, the average value calculation will not be as accurate as when using a digital mAs meter.

What Is Being Calibrated?

During an x-ray calibration, the three parameters that are adjusted by the service engineer are: tube voltage (kVp), tube current (mA), and exposure time. The x-ray service engineer will adjust the kVp, mA, and time so that the values that are set on the x-ray control equal the actual values delivered to the x-ray tube. These three factors are controlled at the x-ray generator control and, consequently, all adjustments are made here, as well.

The amount of kilovoltage that will be applied to the x-ray tube is determined by the primary voltage selected at the x-ray control. This primary voltage (e.g., 100 VAC), when applied to the primary winding of the high voltage transformer, will be stepped up to the desired secondary voltage (e.g., 100 kVp) which is then applied to the x-ray tube. The actual values of primary and the corresponding secondary voltages can be obtained from the ***transformer characteristic graph*** for that particular high voltage transformer.

As stated in Chapter III, the amount of tube current is regulated by a voltage that is applied to the filament transformer. When an mA station is selected on the x-ray control, a corresponding voltage will be applied (during an exposure) to the appropriate filament winding of the x-ray tube. The filament is then heated to the correct temperature to deliver the desired amount of tube current.

The control of both the kVp and the mA, however, is not as straightforward as it may appear. An x-ray calibration would be simplified greatly if a given primary voltage would produce a corresponding kilovoltage that remained constant through the entire range of selected mA's. In addition, the calibration would also be much easier if a given filament voltage would produce a specific tube current regardless of the kilovoltage applied to the x-ray tube.

Unfortunately, the above scenarios do not apply to x-ray tubes in the real world, and, consequently, the calibration of x-ray systems is much more involved. The reason for this has to do with the fact that the tube voltage and tube current interact with each other. This interaction is attributed to two characteristic phenomena of x-ray tubes: ***the x-ray tube loading effect*** and ***the space charge effect.***

An x-ray tube, like all vacuum tubes, does not represent a fixed load to the power source. Initially, the tube will appear highly resistive to the power source. However, once current begins to flow through an x-ray tube, the resistance of the tube decreases in direct proportion to the current flowing through it. If the current is high enough, the resulting low resistance of the tube will cause significant ***loading*** of the power source. This excessive loading, in effect, will "pull down" the supply voltage (i.e., the primary voltage) which then also reduces the kilovoltage that is applied to the tube.

The loading effect is magnified by the inefficiencies inherent in the design of transformers. All transformers experience two types of energy loss termed ***IR losses*** and ***regulation losses.*** These losses are given off as a heat energy and result in a reduction in the output secondary voltages. The service engineer compensates for these transformer losses during calibrations by increasing the primary voltage as the selected mA and, therefore the load is increased.

A ***Transformer Characteristic Curve*** (Figure 24) is available for all high voltage transformers and is usually included with the service literature for each x-ray unit. This curve shows the relationship between the primary voltage and secondary voltage of the transformer under different loading conditions. From the curve the service engineer can determine the actual primary voltages required to achieve a given kVp for each mA selection. From the sample characteristic curve on the following page, it can be seen that at 80 kVp there is a significant difference in primary voltage between 50 mA (120 VAC) and 300 mA (190 VAC). This chart provides valuable information that can be used during the calibration.

The second characteristic of x-ray tubes that affects the calibration relates to the regulation of tube current. When the filament wire inside the x-ray tube is heated, electrons will be "boiled off" of the wire by process called ***thermionic emission.*** As the temperature increases, an "electron cloud" will form around the filament wire. This electron cloud, referred to as the ***space charge,*** will increase in size proportionally with an increase in tile filament current.[19]

---

19. Filament current should never be confused with x-ray tube current, although they are closely related. Filament current, in the range of 1 amp to 6 amps, flows directly through the filament wire and heats the wire to "boil off" electrons. Since the number of electrons liberated dictates the amount of current that will flow through the x-ray tube, filament current controls x-ray tube current. X-ray tube current is in the milliamp range (i.e., 10 mA to 1200 mA).

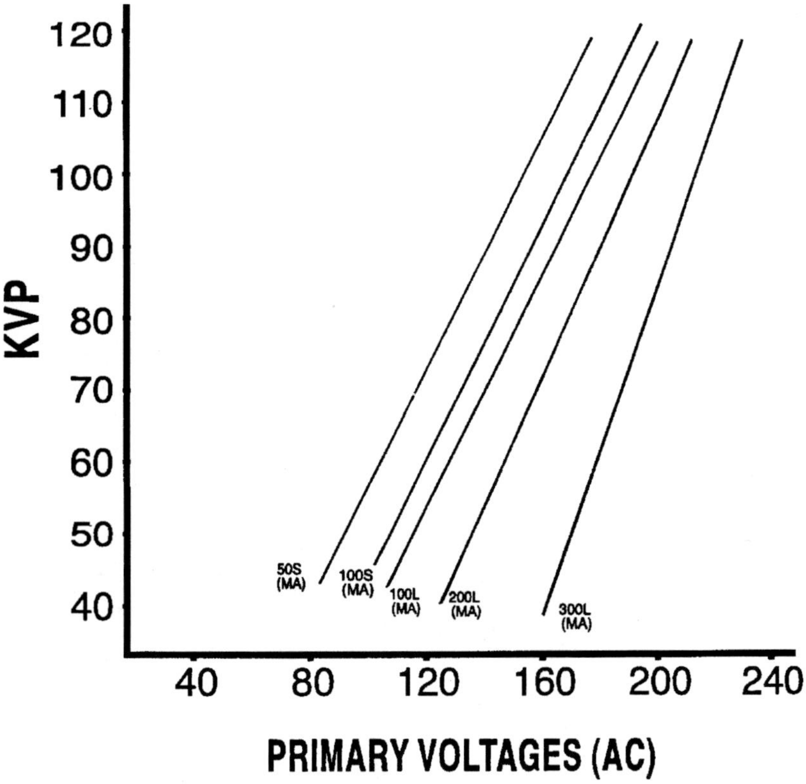

Figure 24. Typical transformer characteristic curve. Because of tube loading, a greater primary voltage is required at higher mA settings for a given kVp setting: 190 VAC for 300 mA verses 120 VAC for 50 mA at 80 kVp.

The space charge phenomena is unique to thermionic vacuum tubes and greatly influences calibrations where high voltages are involved. At higher voltages, the electrons in the cloud become more strongly attracted to the anode (positive electrode) of the x-ray tube. As the kVp of the x-ray tube is increased, more of these electrons in the cloud are "pulled" toward the anode, thus, causing an increase tube current. Specifically, at a given filament voltage (and corresponding space charge) there will be more tube current at 100 kVp than at 60 kVp.

As a result of this phenomenon, all x-ray generators must provide a means to compensate for this space charge effect. ***Space charge compensation*** is provided in all x-ray units and must be adjusted during calibration phase. Space charge compensator circuit effectively lowers the filament voltage at higher kVp's and raises the filament voltage at low kVp settings.

The last parameter adjusted during an x-ray calibration is the exposure time. Because of the high quality of the digital timers provided in today's x-ray units, little adjustment is usually required. In fact, many newer x-ray units do not even provide a means for timer adjustments. As a rule, the timer either works or it doesn't and adjustments are rarely made. Therefore, the service engineer will simply confirm the correct exposure time as measured on the test device.

X-ray Tube Calibration

The procedure for calibrating an x-ray tube is standard for most x-ray equipment on the market today. A general calibration procedure which can be used on most x-ray generators is provided below. By carefully reviewing the entire procedure, the service engineer can better understand what is being accomplished during the x-ray calibration. As always, the service engineer should follow the manufacturer's guidelines when available.

The basic procedure consists of first checking the ***no-load primary voltages*** for each kVp setting. These voltages are usually listed in the service manual but can also be obtained from the transformer characteristic curve described above. Next the service engineer will adjust each mA station so that the mA set at the control is equal to the actual mA read on the test device.

This preliminary mA adjustment is performed on a kVp station that is not affected by the space charge compensation circuits (usually 80 kVp). As a rule, large focus is always calibrated first, followed by calibration of the small focus. Once the preliminary mA adjustments are completed, the space charge compensation circuits are adjusted for each mA station.

Some x-ray generators are designed in such a way that the mA calibration and space charge compensation are combined in one step. This is accomplished by calibrating each mA station at a low kVp point (60 kVp) and at a high kVp point (90 kVp), for example. The low and high kVp points essentially create a slope that is used in the filament control which automatically includes the space charge compensation.

The actual means of adjusting the mA depends on the manufacturer's design. In the more basic generators, adjustment is accomplished by moving a sliding tap on a large wire wound resistor, usually rated at 100 watts, in order to increase or decrease resistance in the filament circuit. Changing the resistance, in effect, changes the voltage applied to the filament transformer thereby altering the filament current.

Some generator designs use potentiometers instead of slider taps to accomplish the mA adjustments. Space charge compensation can also be accomplished by a resistor adjustment, or by selecting different taps on a special boost/buck transformer. The latest generator designs use digital control for kVp and mA regulation, and adjustment is accomplished by using potentiometers or by soft key control at the console.

With mA calibration completed, the kVp's are adjusted for each mA station. Adjusting the kVp is also accomplished by sliding taps on resistors, potentiometer adjustments, or through software control. Some single-phase generators, however, do not provide a fine adjustment for the kVp. Since the kVp and mA closely interact with each other, these generators are calibrated by setting the line tap on the autotransformer first, and then adjusting the mA's so that both the kVp and mA accuracy is achieved.

With single-phase generators that have an analog kVp panel meter and provide kVp adjustment, what is actually being adjusted is the kVp prereading of the kVp meter. An 80 kVp reading on the panel meter, for example, is simply a representation of the corresponding primary voltage selected by the kVp major and minor tap switches. Here the goal of kVp calibration is to adjust the kVp resistor to cause the indication on the display to match the actual kVp delivered to the x-ray tube.

When calibrating the kVp on three-phase generators, the service engineer will also adjust the primary voltages, but the adjustment is usually done by electronic means. With most autotransformers in three-phase generators, the "wiper" that selects the primary voltage is motor driven. The motor control circuitry utilizes *feedback signals* that monitor the line voltage, as well as, the actual kVp delivered to the x-ray tube for precise regulation. This method provides very accurate kVp control. Usually the incoming line voltage is also regulated by this method to help maintain accurate kVp regulation.

The kVp adjustments for three-phase generators are made by adjusting a potentiometer that controls the servo drive circuit, which then moves the wiper of the "auto" to obtain the correct primary voltage for the selected kVp. On newer designs and high frequency generators, the adjustment is made through the soft keys.

As said, the timer circuits of today require very little adjustment. Although some analog timers are still in existence, most timers today are digital and are very accurate. If a timer adjustment is needed, it is usually accomplished by adjusting the clock speed of an oscillator on the timer circuit board. Analog timers usually require potentiometer adjustment for each time setting. Newer software-controlled units allow the engineer to make the adjustment at the console.

A final note concerning calibration should be mentioned. While the engineer is performing x-ray calibration, he or she should monitor the kVp and mA waveforms to ensure correct operation of the x-ray generator. In addition, the *engineer should also monitor the line voltage supplied to the generator.* A significant drop in the line voltage during an exposure would indicate a deficiency in the power line supplied to the x-ray generator. This step is crucial to the performance and longevity of the radiographic equipment.

The manufacturer will only guarantee the output accuracy of the x-ray generator if the ***line regulation*** is within their specifications. Consequently, the line regulation must be measured and recorded in the installation manual. The measurement should be made after the completion of the x-ray calibration, since line regulation is tested at near maximum generator output. Most manufacturers of radiographic equipment require a 5 percent line regulation.

The line voltage is first measured while the machine is idling in "standby mode" (no-load). A second measurement of the line is made during an exposure at (or near) maximum generator technique (load). The percent of line regulation can be determined by using the formula: % regulation = $(V^{no\text{-}load} - V^{load})/V^{load}$.

X-ray Calibration

CONFIRM LINE VOLTAGE AND MECHANICALLY ZERO THE KVP METER BEFORE STARTING THIS PROCEDURE!

1. Insert the calibration test equipment as described earlier in this chapter.
2. Check no-load primary voltages
3. Perform tube seasoning and/or warm-up if needed.
4. Take several exposures at different kVp settings and confirm anode and cathode waveform (i.e., balancing).

5. Remove cathode side of divider to eliminate loading effects. Calibrate using anode waveform only.
6. Select 80 kVp, 1/10 S (100 ms), and begin with the lowest mA station on large focus.
7. Adjust all mA stations on large focus, then adjust small focus, keeping kVp and time constant. *To avoid overheating the x-ray tube, wait at least 15 seconds between exposures!*
8. Select a large focus setting and check the space charge adjustment at 60 kVp and 120 kVp and adjust if necessary. Repeat space charge compensation check for the remaining large focus stations and for all small focus stations.
9. Perform the kVp adjustment, if applicable, for each individual mA station.
10. Recheck and fine tune mA's.
11. Check the timer and adjust if necessary.
12. Measure and record the line regulation.

With the tube calibration complete, this is a good time to measure the focal spot size of the x-ray tube. Although most x-ray tube manufacturers perform focal spot measurements prior to shipping, they should be measured during installation so that there is a record to keep with the equipment file.

For this measurement, a ***lead test star*** is commonly used. The test star is positioned in the x-ray field, exposed on a test film, and then measured. When the test star is radiographed, it produces a "spoked wheel" image as seen in Figure 25. The radiation from the actual focal spot will create a blurred region in the spoked pattern. The blurring is due to the penumbra of the focal spot. By measuring the length and width of the blurred region, the actual focal spot size can be determined using the formula:

$$F = N/57.3 \times D/(M - 1)$$

where: F is the focal spot size.
N is the angle of the star pattern (written on the pattern)
D is the diameter of the blurred region (L or W)
M is the magnification

General instructions for using the test star are listed below:

1. Position the star in the center of the x-ray field, perpendicular to the central ray. It should be positioned at a distance to achieve a 2×

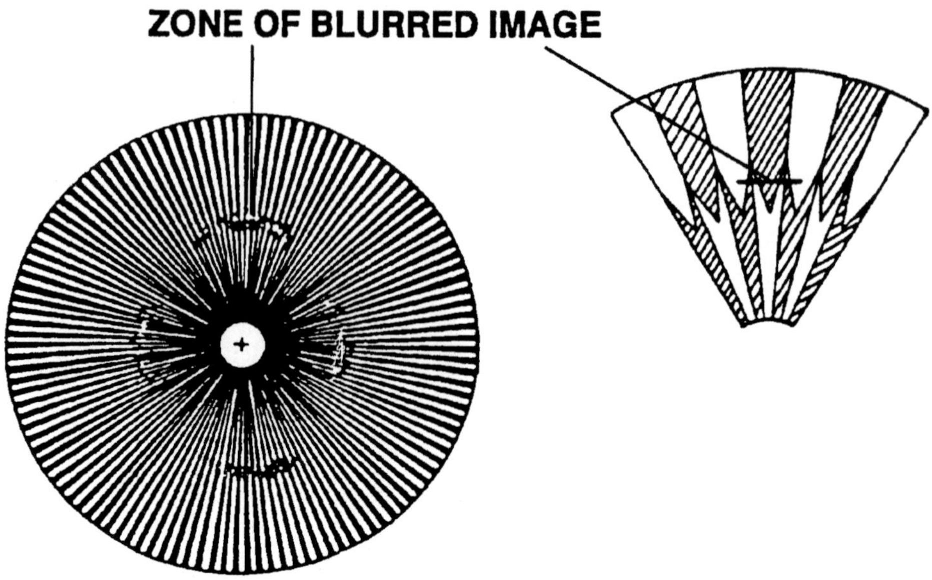

Figure 25. Lead test star image. Measurements of the blurred region of the image are used to calculate the focal spot size of the x-ray tube.

magnification (i.e., equidistant-distant from the focal spot and the film). It is convenient to tape the star to the collimator on an overhead tube.

2. Load film into a cardboard holder (no screen cassettes) and position carefully in the x-ray field.

3. Use an average technique to achieve approximately a 1.5 film density (i.e., 75 kVp, 100 mA, adjust time).

4. Develop film and measure the diameter of the image. Divide the diameter of the image by the diameter of the actual star. This gives the magnification factor (M).

5. Find the outermost point where the spoked pattern just begins to blur, and mark. Do this for length and width.

6. Measure the distance between the marks and use the formula above. Do separate calculations for length and width. Verify accuracy.

Document the results along with the technique factors and test set up geometry (i.e., SID and OID) and file this information with the installation report. This information can be used as a reference when image detail begins to degrade.

Collimator Adjustment and Alignment

When installing a collimator, the service engineer must perform several operations. For manual collimators:

1. The intensity of the light field must be verified.
2. The light field must be made to match the x-ray field.
3. The cross-hair window must be aligned.
4. The bucky centering light must be aligned.
5. The SID indicator must be adjusted.
6. The field size indicators must be verified.
7. The correct amount of filtration must be verified.

For automatic collimators, all of the above steps must be performed in addition to the adjustments of the PBL system.

Great care should be taken during this phase of the installation since it is critical that these adjustments are performed correctly. If the collimator is not set up properly, the patient and x-ray personnel could receive significantly more radiation than is necessary. Also, if the light field and x-ray field are not properly aligned, certain anatomy of the patient may be cut-off of the film. When this occurs, the technician must "retake" the film over again which not only increases the dose delivered to the patient, but also frustrates the technicians and radiologist.

A setup procedure for manual collimators is provided below, followed by the additional steps needed to setup an automatic collimator. The engineer should keep in mind the fact that all automatic collimators have an override switch to place the collimator in "manual mode" of operation. Consequently, all automatic collimators must meet the same guidelines as for manual collimators.

Light Field Intensity

An important function of the collimator is to produce a light field that accurately defines the x-ray field. Until very recently, there were no specific requirements for the light output level of the field lamp and, as a result, the designs of lamp circuits in collimators vary among the manufacturers. The light output levels likewise vary for the different brands collimators.

Because of this wide discrepancy in light output, a new standard for collimators was created: if a "light localizer" is to be used to indicate the x-ray field, it must have a ***minimum intensity level*** so that the light field produced will be clearly visible in normal room lighting. The new guideline states that: *The minimum average illumination should be not less than 160 lux (15 foot-candles) at 100 cm or maximum SID, whichever is less.* Collimator field lamp output is measured with a light meter (or ***photometer***) that is placed at the center of the light field.

Many older collimators (i.e., those manufactured in the late 1970s or early 1980s) will not meet this minimum requirement. Moreover, there is usually little or no adjustment possible to increase the light output of the field lamp. With some designs, it is possible to select a higher tap on the transformer that supplies the voltage to the lamp. Increasing the supply voltage will cause the lamp brightness to increase but will significantly reduce the life of the lamp. As a rule, *any modifications should always be preapproved by the manufacturer before proceeding with the modification!*

If the lamp output cannot be adjusted, the collimator will have to be replaced or refitted with a new lamp assembly and lamp control circuitry. Fortunately, all newly manufactured collimators will easily meet this minimum requirement. The installer should nonetheless measure the light output level and record that value in the service manual for future reference.

Light to X-ray Field Congruence Test

As said, one of the important functions of the collimator is to project a light field onto the film plane that gives a visual indication of the x-ray field. The edges of the light field must therefore match the edges of the x-ray field to within specified limits. The procedure for testing this congruence is described below.

1. Position the tube 40″ above the tabletop and place a film cassette on the table. Tube should be at 0° angulation.
2. Center the cassette to the light field and close the collimator shutters down to approximately the next smaller film size.
3. Mark the edges of the light field with some type of metal marker that will produce a sharp border on the film.
4. Expose the film to achieve a density of 1.0 and then process the test film.

5. Measure the difference between the edges of the light field (as indicated by the image of the markers) and x-ray field.
6. The sum of the difference for the long axis and the sum of the difference for the cross axis should not exceed 2 percent of the SID.
7. Adjust the light field if necessary and repeat this procedure until it is within the specified limits.

The two directions that the lamp can be adjusted are in the long direction and the cross direction as referenced to the tabletop. Generally, the lamp bracket which holds the lamp in place is moved to correct errors in the long direction. The field lamp mirror, which is found in all collimators, is adjusted to correct errors in cross direction. If the mirror assembly is the stationary type, the cross adjustment is also made with the lamp bracket.

Cross-Hair Window Adjustment

This adjustment is for centering the cross-hair pattern that is projected from the collimator onto the film plane. The cross-hair pattern is used to precisely align the x-ray tube to the patient and receptor. The plastic window, located in the bottom of the collimator, is "etched" with a cross-hair pattern. When the collimator lamp is lit, the cross-hairs are projected to the film plane.

The cross-hair window adjustment is easily accomplished by loosening the screws that hold the plastic window in place and centering the cross-hair pattern within the light field. A quick method for checking if the cross-hair is properly centered is to close the collimator blades down slowly, one set of blades at a time, until a narrow line of light is projected. This narrow light line should be positioned at the center of the cross-hair. Centering can also be verified be placing a film cassette on the tabletop and marking the cross-hair center with a lead bead, and then exposing the film.

Bucky Centering Light Adjustment

All collimators have a bucky centering light that allows the technician to properly align the x-ray tube to the cassette tray, thereby ensuring full film coverage. The light source for bucky centering is usually derived from the collimator field lamp and is projected to the front of the bucky assembly by the use of a prism or mirror. By adjusting the prism/mirror assembly, the light-line can be centered to the bucky. The service engineer should always check the adjustment by exposing a film with the collimator blades manually closed down to a size that is slightly smaller than the actual film size and then checking to see that the exposed area is centered on the film.

SID Indicator Set-Up

Most collimators provide some means to identify the ***Source to Image Distance*** or ***SID.*** The SID is the actual measurement of the distance from the focal spot of the x-ray tube to the film plane. The indicated SID must be accurate since the x-ray technician will use specific SID's to obtain the desired film coverage and sharpness required for the particular exam. Also, a specific SID must be used when a ***focused grid*** is used. In fact, the exact SID that must be used for each focused grid is clearly indicated on the grid label. Using an SID different from that stated on the grid will result in grid lines on the radiograph (see Chapter VIII). In addition, if the SID indication is not accurate, the ***field size indicator*** on the collimator face will not be accurate and, therefore, cannot be used.

In the most basic collimator design, the device used for SID indication is usually a metal tape measure that attaches directly to the collimator housing. The tape measure has been modified to account for the focal spot to collimator distance, and therefore, the graduations on the tape do not begin at zero. The x-ray technician can obtain accurate SID readings using this modified tape.

Some collimators use a lamp or numerical display to indicate the SID. The SID lamp will "light up" when the tube is positioned at the correct SID. More sophisticated SID systems continuously monitor and display the SID on a digital readout.

If the SID indicator is not included with the collimator, it is usually provided with the tube support assembly. Here the SID indication will consist of a graduated scale that ranges from 36 inches to 72 inches. This SID scale attaches directly to the vertical tube column to indicate vertical SID. A second graduated scale is attached along the ceiling rail to indicate horizontal SID. If an automatic collimator is being installed, interlock switches will be positioned at the 40 and 72 inch SID marks to satisfy the collimator logic.

In order to set up the SID indication properly, the x-ray service engineer must know the exact location of the focal spot within the x-ray tube housing. The information regarding the focal spot location (with respect to the housing port) can be obtained from the manufacturer of the x-ray tube. In fact, this measurement should have been used when the spacers were added at the time the collimator was mounted to the tube. Some tube manufacturers provide a mark on the end cap of the x-ray tube to indicate the focal spot location. This mark, however, should not be used until it has been verified by actual measurements.

To Set Up the SID Indicators

1. From the tube data sheet, obtain the focal spot to x-ray tube port distance (where the collimator upper swivel is attached). Carefully measure and place a mark on the tube housing or end cap to indicate the location of focal spot. If the tube manufacturer already provides a mark, verify that it is correct.
2. Next, move the tube away from the tabletop until a distance of 40″ from the film plane to focal spot mark is achieved. Be sure to lock the tube in place.
3. If the collimator is equipped with a modified tape measure and if the correct number of spacers were added when the collimator was mounted, the tape measure should now also indicate 40″ to the film plane.
4. If the collimator utilizes a SID switch to activate a lamp indicator or interlock, mount the switch and switch activator (for 40″ SID) on the tube column at this time.
5. If there is no SID indication provided on the collimator, then a graduated scale should be placed on the tube column to indicate 40" SID to receptor. Some manufacturers of tube supports provide a dual reading scale that indicates 40″ to table top, as well.

Once the vertical SID indication is verified, the horizontal SID indicators can now be set by using the same mark on the end cap of the tube, which is known to be accurate. The SID setup procedure should be performed as described above for both 40″ and 72″ horizontal SID's. The installer should remember to make all the measurements from film plane in the wall receptor to the focal spot of the x-ray tube.

Field Size Indication

All collimators are required to provide an indication of the x-ray field size for the SID that is being used. The field size indicators are graduated scales that are mechanically attached to the shutter control knobs so that the size of the x-ray field can be directly read as the knobs are turned.

The accuracy of the field size indicators can be verified by performing a simple test which is detailed below. For this test, the service engineer will set a specific field size by adjusting the shutter control knobs while reading directly from the scale on the collimator. He or she will then expose the film and measure the actual field size produced on the film.

To Check Field Size Indicator

1. Position the x-ray tube for 40″ SID to the table top (make sure SID was verified).
2. Place a 10 × 12 inch (24 × 30 cm) cassette on the table and center to the x-ray tube.
3. Close shutters down until the indicator reads 8 × 8 inches (15 × 15 cm) and expose film to a density of at least 1.0 D.
4. Measure the size of the exposed area on the film and verify that both the length and width are within 2% of the set SID (i.e., 2% of 40 inches or 0.80 inches).

If there is a deviation of more than 2 percent of the SID, the indicators must be adjusted. Adjustment is usually accomplished by loosening the set screws that hold the scales to the longitudinal and horizontal shutter knobs and then rotating the scales to read the actual field size obtained from the film. Once an adjustment is made, the field size testing procedure described above must be repeated.

Automatic Collimators

Once all of the adjustments are completed for manual operation, calibration of automatic mode can begin. An automatic collimator requires several external signals to function properly, and therefore, there are many more interconnecting wires needed for automatic operation. Whereas a manual collimator requires just one three-conductor cable to provide power for the lamp circuit, an automatic collimator requires:

1. A cable for horizontal SID sensing.
2. A cable for vertical SID sensing.
3. A cable interfaced with the x-ray generator for the exposure interlock circuit.
4. A cable interfaced to the table receptor for cassette size sensing.
5. A cable interfaced to the wall receptor for cassette size sensing.
6. A cable connected to the table tilt monitor (if used).
7. A cable for stereo/tomo interlock.
8. A power cable.
9. An interconnecting cable between the collimator control and the collimator head.

The service engineer should verify that these cables are connected before proceeding with the collimator adjustments.

In automatic or PBL mode, a reference signal representing cassette size is derived directly from the cassette tray, and is fed back to the control logic of the collimator. A second signal representing the actual shutter field size is also sent from the collimator to the control logic where it is then compared to the reference signal. If the actual shutter size does not match the cassette size, the exposure will be blocked by the control logic until such time as it does.

Semiautomatic collimators do not have shutter drive motors and require the technician to turn the shutter knobs manually until the logic is satisfied (indicated by a "ready" light). With fully automatic collimators, the control logic sends a "drive" signal to the shutter motors to open or close the shutters until the logic is satisfied. The "drive" signal, however, is only active if the tube is at the proper SID. If not, the exposure will be inhibited until the tube is positioned correctly (usually indicated by a lamp on the front of the collimator). Also, some collimators will inhibit an exposure if the filter is not in position in the collimator.

The collimator logic monitors several other parameters besides cassette size, shutter position, and SID. The logic also monitors the table angle if it is a tilting model, the rotation of the x-ray tube, the presence of a film cassette in the cassette tray, and the mode of operation that is selected at generator control (i.e., tomo or radiographic). If any of the above conditions does not satisfy the control logic, an exposure hold condition will result.

Automatic Collimator Calibration

Although automatic collimator design varies among the different manufacturers, the goal of the collimator calibration is the same for all collimators. During the calibration procedure the engineer will adjust the cassette trays to output standard signals for each different cassette size. The output signals are sent to the control logic where they are compared to precisely adjusted reference voltages. The control logic is then adjusted to output the correct motor drive signals to the collimator shutter drive circuits. Finally, all the collimator exposure interlocks are connected and tested.

The size sensing cassette tray outputs a specific resistance when a cassette is placed in the tray. Every cassette tray manufacturer uses the same standardized values for output resistances, which makes setup much easier. Listed below are the standard outputs for each cassette size.

Cassette Size	*Output (in ohms)*
5″ (12 cm)	60
7″ (18 cm)	207
8″ (20 cm)	280
10″ (25 cm)	427
11″ (28 cm)	500
12″ (30 cm)	573
14″ (35 cm)	720
17″ (43 cm)	940

With manual cassette trays, a film cassette is placed in the tray and then locked in place in the lateral or cross-table direction by a front and rear clamp. These clamps are mechanically attached to gears that engage a potentiometer. The potentiometer outputs a resistance that corresponds to the size of the cassette for the lateral direction.

When the cassette tray is inserted into the receptor cabinet, an ***actuator pin,*** mounted inside the receptor cabinet, pushes the spring-loaded arm of the cassette tray into position against the side of the cassette. A feedback potentiometer for the "long" direction is mechanically attached to this arm. If there is no cassette present in the tray, this side arm will continue to move towards the center of the tray and activate a switch that will inhibit the output signals from the cassette tray. Cassette tray calibration is accomplished by adjusting the output resistance of the two potentiometers and adjusting the cassette sensing switch.

Size-Sensing Cassette Tray Calibration

1. Place the size sensing cassette tray on tabletop and remove the cover to expose the adjustment pots.
2. Place an 8 × 10 inch cassette in the tray and clamp into place, making sure that the cassette is centered.
3. Move the side arm into position by applying pressure to the *heel* of the arm (not the end of the arm). Move the arm until it rests firmly against the side of the cassette and then wedge it into place with a screwdriver. If a clamping mechanism is used to sense the size of the long dimension of the cassette, position it so that it rests firmly against the side of the cassette.
4. Verify the correct resistance between pins #1 & #5 of the cassette plug for cross direction and between pins #4 & #8 for the long direction. Refer to the table on the previous page. Adjust if necessary.
5. Rotate the cassette 90° in the tray and verify that the resistances are accurate for this cassette orientation.
6. Replace the cassette with a 14 × 17 inch cassette and check the readings in both directions. Adjust if needed.
7. Remove the cassette and move side arm (or clamp) in towards the center of the cassette and verify that the "cassette present" (auxiliary) switch is activated at the point just before the end of the arm reaches the edge of the center opening of the cassette tray.
8. Repeat this procedure for the wall receptor cassette.

Some manufacturers offer an automatic style bucky. With an automatic bucky there is no need for a cassette tray. Auto-loading buckys are completely motorized and will automatically position the cassette when it is inserted into the bucky. When the cassette is in the exposure position, the cassette size is sensed and the correct resistance is selected from a resistor network. No sizing adjustment is necessary for the automatic bucky.

Installing the SID Switches

At this point in the automatic collimator adjustment procedure, the control logic reference voltages should have been adjusted and the cassette trays calibrated. In the next step, the SID switches will be mounted and adjusted. This is a simple process, since the focal spot of the x-ray tube has already been established and is indicated with a mark on the end cap of the x-ray tube. The SID switch is mounted in such a way that when the tube is at the desired SID, the switch will be activated and the associated interlock circuitry of the collimator will be enabled.

For the vertical SID, the x-ray tube should be positioned so that the focal spot is 40″ above the film plane in the table receptor and then locked into place. *Make sure the radiographic table is in the exposure position if installing a hi-lo radiographic table.* The SID switch should be mounted to the x-ray tube support on a section of the tube support that is stationary. This is done so that the wires that attach to the switch can run internally along the tube support where there is no chance that they will become detached when moving the x-ray tube.

A ***switch detent*** is used to activate the SID switch. The detent must be positioned on the moving part of the tube support in such a way that, when the tube is at 40″ vertical SID, the switch should be directly opposite the detent. The SID switch is actually a microswitch with a roller style actuator and is activated when it rolls onto the detent. When the switch is engaged, a logic signal is sent to the collimator control to indicate that the tube is positioned at 40″ SID.

Once the detent is mounted, the engineer should verify that the switch does in fact activate at 40″ SID by moving the x-ray tube up and down and observing the switch. If it is working properly, the 40″ vertical SID wires can be connected to the collimator control logic. This procedure should be repeated for any other vertical SID's that will be used and to mount the horizontal SID switches.

For the horizontal SID, a switch for 40″ SID and one for 72″ SID should be installed (36″ and 48″ SID's are rarely used today). The two SID switches should be mounted to the ceiling rail of the tube support at the premarked location. Again, the switches are mounted to a stationary surface so that the associated SID wires will remain fixed in place, and consequently, there will be no chance that they will be pulled from the switch. Next, the detents should be mounted directly to the moving tube support so that the switch will be activated at the correct SID distance. When the switch and detents are mounted, the tube support should be moved along the rail while the engineer verifies proper operation of the SID switches.

Continuous Vertical SID

Some automatic collimator designs incorporate a ***continuous vertical SID monitor.*** The SID monitor continually tracks and numerically displays the vertical SID for the entire range of tube travel. The continuous vertical SID option is considered an upgrade and is used in place of the vertical SID switches. Consequently, there is no need to calibrate the 40″ vertical SID switch.

The setup procedure for continuous vertical SID involves setting a reference voltage at 40″ SID in the control logic and then making adjustments for proper tracking over the 36″–72″ range of SID. Also, upper and lower SID limits are set so that the exposure will be inhibited when these limits are exceeded. The service engineer should follow the specific guidelines written by the manufacturer for setting up the continuous SID monitor.

Automatic Collimator Operational Checkout

The accuracy of the automatic collimation system must be checked for all cassette sizes and SID's that are going to be used. The following test is used by many radiation physicists and will confirm all functions of the PBL system. In this procedure, one cassette is exposed several times to verify all the cassette sizes.

1. Set the x-ray tube at 40″ SID, 0° angulation, and energize all the locks.
2. With the collimator switched to automatic mode, insert an 8 × 10 inch cassette into the receptor.
3. The shutters should drive to the correct field size and then stop. Verify the correct field size as indicated on the collimator. The collimator should be in an exposure hold mode while the shutters are moving.
4. Switch the collimator to manual (bypass) mode.
5. Remove the cassette and insert a 14 × 17 inch cassette. Take an exposure that would achieve approximately a 0.5 density. *Do not Develop.*
6. Remove the 14 × 17 cassette and set it aside.
7. Switch to PBL mode again and insert a 10 × 12 inch cassette. When the shutters stop moving, switch back to manual mode and verify the correct field size indication on collimator.
8. Remove 10 × 12 cassette and insert the same 14 × 17 inch cassette used in step #5. Take an exposure using the same technique as in step #5.
9. Remove the 14 × 17 cassette and set aside.
10. Repeat the procedure for an 11 × 14 inch cassette.

11. Develop the film. Each film size should be distinctly outlined on the film. If not, repeat steps 1–11 using a lighter technique.

12. Measure the radiated area for each size and record the results. The length or width for each size should be within 3 percent of the SID. Three percent of 40″ yields +/–1.2″.

13. Repeat entire procedure for the wall receptor at 40″ and 72″ SID.

14. In PBL mode, confirm that no exposure is possible if the tube is tilted more than 3° from the film axis. If tilting table was installed, verify the fact that if the table is tilted more than 3° from the tube axis, no exposure is possible.

If any results from the above procedure are out of tolerance, the automatic collimation system will have to be recalibrated. To isolate the area of the problem, see if the percent error is the same for both the table and wall bucky (if both are used). If both of the cassette trays are out of tolerance this would indicate that the collimator reference and/or tracking adjustment is not correct. Also, check that the field size indication on the collimator agrees with the actual measured field size. If they agree, check the resistance output of each cassette tray.

Automatic Exposure Control

An Automatic Exposure Control system, or AEC for short, is available as an option with most x-ray generators. An x-ray generator equipped with the AEC option can produce films that are more consistent in quality than one that provides manual techniques only. Also, since fewer "retakes" are performed in a room equipped with AEC most x-ray technicians, are well as most radiologists, request that their equipment include the AEC option.

The AEC unit senses the x-ray radiation delivered to the patient and then automatically terminates the exposure when the optimum film density is achieved. Most AEC units can accommodate several different film/screen combinations and usually provide a means of film density compensation. A properly calibrated AEC unit will consistently reproduce optimum film densities regardless of the size of the patient.

When using AEC, the x-ray technician simply selects a kVp and mA setting as is normally done for manual exposures. However, since the exposure time is now controlled by the AEC unit, the technician will instead select a ***backup time*** on the control panel. A backup time is required as a safety measure so that if the AEC fails to function properly, the x-ray unit will then terminate the exposure. The backup timer is usually set to a time that is much longer than the anticipated exposure time (i.e., at 1–2 seconds).

An AEC unit consists of three major components. The ***detector*** or ***pickup,*** located in the image receptor assembly, senses the x-ray radiation that is delivered to the patient. The detector is connected through a shielded cable to a ***control unit*** which converts the detector output signal to an exposure stop signal that is then sent to the x-ray generator. Finally, the AEC unit (with the exception of the solid-state type discussed below), must have a high quality ***regulated power supply*** in order to achieve consistent exposure control.

The Detector

Since the detector assembly is used to sense radiation, it must be placed in the path of the x-ray beam, usually directly in front of the film cassette in the image receptor. The function of the detector is to sense x-ray radiation that has passed through the patient and convert that radiation energy to an electrical signal that is sent to the control unit where it will be used to control the exposure time. Currently, there are three different types of AEC units in use today and these are distinguished by the type of detector employed.

If the detector contains one or more ***photomultiplier tubes,*** the AEC unit is referred to as a ***phototimer.*** The phototimer was the earliest type of AEC unit, introduced by Paul Hodges, its designer, in 1942. It had such an impact on radiography and was so widely used that, today, service engineers often refer to any AEC unit as a phototimer.

The phototimer detector assembly (sometime referred to as a paddle) consists of a fluorescent screen encased in a light-tight seal. The purpose of the fluorescent screen is to convert x-ray radiation to light energy. A "window" is provided at one edge of the detector to allow the light to exit. A photomultiplier tube, or ***PMT,*** is optically connected to the paddle at the exit window and will convert the light energy to an electrical current signal which is amplified and sent back to the AEC control for further processing.

The PMT is a special type of vacuum amplifier tube that consists of a light sensitive ***photocathode,*** a series of ***dynodes,*** and an ***anode*** (see Figure 26). The photocathode converts light energy into a stream of

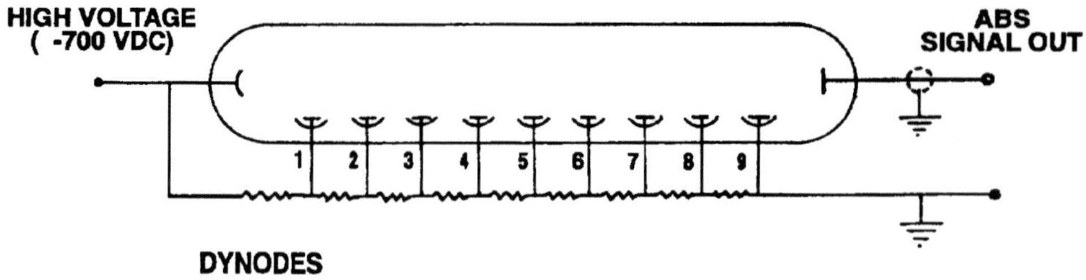

Figure 26. Photomultiplier tube (PMT). The dynodes are wired in series to achieve high amplification.

electrons that will then strike the first dynode. A dynode is a special electrode that is used for signal amplification. In essence, if one electron strikes the surface of a dynode, that dynode will emit 5–7 secondary electrons. Consequently, by using 9 (or more) dynodes in succession, very high amplification can be achieved. The resting current signal is sent from the anode of the PMT to the AEC control circuitry.

The PMT is commonly used in the exposure control circuits of x-ray equipment because it provides very reliable, trouble-free operation. They do, however, require a fairly high supply voltage (600 VDC to 1000 VDC) for operation. In addition, this power source must provide a regulated, noise-free output to achieve the optimum performance needed for consistent exposure control. During the AEC calibration, the engineer will set the gain of the PMT by adjusting the amount of high voltage is applied to it.

The second type of AEC detector utilizes an ***ion chamber*** instead of a PMT for signal conversion. The ion chamber is contained within the detector assembly and is also positioned directly in front of the film in the receptor as in the case of the phototimer detector. Instead of performing light conversion, however, this detector utilizes the ionizing effects of x-ray radiation. When x-rays strike the ion chamber, they will ionize the gas within, producing an electrical potential across the chamber.

Usually, a preamplifier is physically attached to the detector assembly. The preamp provides the initial signal conditioning required before it is sent to the AEC control for further processing. Ion chambers require a regulated power supply voltage in the range of 300 VDC to 400 VDC. Generally, adjustments are provided in the preamplifier for balancing the outputs of each ion chamber. Otherwise, all other adjustments are performed in the amplifier section in the AEC control.

The third type of AEC unit uses a ***solid state detector.*** A solid state detector is an exit-sensing device that is usually positioned in the x-ray beam, just behind the film. The sensor detects x-ray radiation that has passed through the patient and film cassette and sends a corresponding electrical signal to the AEC control unit.

The operating principle of the solid state detector is very basic. If a diode, for example, is placed in the path of the x-ray beam, it will conduct current. The amount of current produced by the diode is proportional to the amount of radiation that strikes it. The solid state detector utilizes this same principle for operation. It is virtually self-powered.

The advantages of this type of detector are its compact size and extremely rugged design. Most important, however, is that it is the only AEC unit that does not require a power supply to bias the detector. The small size of the solid state detector makes it ideal for use in mammography equipment.

AEC Control

As stated, radiation, when striking the detector, is converted into a current signal. The output signal from the detector is integrated at the ***AEC control.*** After initial processing, the signal is converted to an analog voltage ramp that is then fed to a comparator in the control. Here the level of the voltage ramp is compared with a reference voltage. With a ramp signal, the voltage level will steadily increase during the exposure. When the level of the ramp matches the calibrated reference signal, an exposure stop signal is generated and sent to the x-ray control.

The actual value of the reference signal is adjusted during the AEC calibration and is dependent on factors that are selected at the operator control. These include: the kVp setting, the film density compensation selected, and the film/screen type. These parameters must be taken into consideration when the engineer performs the AEC calibration.

The AEC control also provides regulated power to the detector (PMT and ion chamber). The supply voltages are very critical to the performance of the detector and therefore the power supplies must be of high quality and very reliable. The optimum operating voltage for the photomultiplier tube is 700 VDC, but can range between 600 VDC and 1000 VDC. Outside the specified operating range, the photomultiplier tube becomes nonlinear and will also have a *reduced signal to noise ratio.* The engineer will adjust this voltage during the calibration.

Ion chambers, on the other hand, require a much lower voltage in the 300–400 VDC range. This voltage, however, is usually fixed and therefore will not be adjusted during the calibration. The engineer need only verify that it is at the correct level.

AEC Calibration

AEC calibration should be one of the last tasks performed on an installation, for several reasons. First of all, the x-ray tube must be aligned and calibrated, since the response of the AEC detector is very much dependent on the selected kVp and on positioning of the x-ray field. Second, to calibrate the AEC properly, the service engineer will need to use the exact type of film and cassettes that will be used in the new room. Many times, the cassettes and film that will be used in the new room are not delivered to the site until the room is complete.

Finally, and most importantly, *the film processor must be operational and stabilized before the AEC calibration can be performed.* The film processor should be running for several days (minimum) so that quality control testing can be performed on the processor to confirm that the temperature and chemistry has in fact stabilized. If any film processor problems are encountered, they must be corrected before proceeding with this calibration. This point cannot be stressed enough.

The goal of AEC calibration is to achieve a desired film density (optimally 1.2 D) for the particular film/screen combination being used. The density should remain constant over the entire kVp range and patient sizes. Test equipment needed for AEC calibration consists of a ***homogenous radiographic phantom*** (usually acrylic) and a film densitometer.[20] The general procedure for calibrating a typical AEC unit is described on the following pages. As usual, follow the manufactures procedure when available.

---

20. Phantoms can be purchased from many x-ray accessory and film suppliers. If a phantom is unavailable, a plastic water bucket filled with 8″ of water can be used to simulate the density of a normal size adult. Also, acrylic can be purchased locally and is commonly used. Buy 1″ thick sheets and cut into a dimension that will be easy to handle and still cover all three of the fields (e.g., 12″ × 12″). 12 acrylic plates should be adequate (1″ of water equals 1″ of acrylic).

The radiographic phantom is used to simulate the density of an average patient. Phantoms are used because of the numerous exposures that will be taken during the AEC calibration. Although the phantom closely mimics human tissue, it cannot exactly produce the effects of an actual patient. Consequently, the service engineer must return to the facility when the first patients are being filmed to "fine tune" the density adjustments.

Field Selection Verification (Three Field Detector)

1. Install test equipment to accurately measure the exposure time if the actual exposure time is not displayed on control front panel. If actual time is displayed, accuracy should have been verified during the x-ray tube calibration.
2. Cover all three fields with a lead blocker.
3. Select the ***left detector field*** at the console, take an exposure, and verify that the actual exposure time is equal to the time set on the backup timer.
4. Remove the lead blocker from the left field only.
5. Take another exposure and verify that the exposure terminates immediately (8.3 milliseconds or less) with the left field selected.
6. If the exposure terminates immediately, then the left field is hardwired correctly and the same test can now be performed on the right field and then on the center field.
7. If the generator goes into backup time while testing the other fields, the detector must be rewired until the field selection switches on the generator control correlate with the field that is energized.

After rewiring has been completed repeat the test for each individual field.

AEC Offset Adjustment

As stated earlier in this section, all AEC devices obtain a signal that represents the amount of radiation delivered to the cassette. That signal is then sent to the AEC control unit where it is processed through several amplifier stages. The ***offset*** of these amplifiers must be adjusted before proceeding with the calibration. Since manufacturers' design of AEC circuits varies greatly, refer to schematic or manufacturers instructions for offset adjustments.

AEC Gain Adjustment

1. Align x-ray tube to the image receptor @ 40″ SID.
2. Place a flat field (homogeneous) adult phantom over entire detector making sure that the three fields are completely covered.
3. Select 80 kVp, 200 mA, ***manual mode*** (AEC off).
4. Take an exposure that will produce a density of 1.2 D, as measured with a calibrated densitometer, and record the exposure time *(note: make sure exposure time is greater than 30 ms to eliminate effect of short time response of the PMT!)*. Several exposures may be required.
5. Switch to AEC mode and select the ***center field,*** normal density (0 density compensation), and the correct film speed (i.e., low, med., or high) for film being used in the room. Place an empty cassette in the tray to account for backscatter radiation.
6. Adjust the gain of the detector until the actual exposure time obtained with the phantom is equal to the time recorded in step #4.

Make sure individual gain pots for each field are at the midpoint position before starting the gain adjustment. With the phototimer, the gain is adjusted by varying the high voltage supplied to the PMT. For the ion chamber, the gain is first adjusted in the preamplifier of the detector, and then a fine adjustment is provided at the integrator in the control.

Balancing the Three Detector Fields

In a properly calibrated AEC unit, each individual field of the detector must produce the same density for a given exposure and patient thickness. The service engineer must adjust the left and right fields to match the center field (calibrated above).

1. Select the left detector field and adjust gain potentiometer for that specific field to match the exposure time obtained for the center field.
2. Repeat this procedure for the right detector field.

KVp Compensation

Because the absorption characteristics of radiation change at different kVp settings, kVp compensation must be performed on all AEC units. This procedure ensures consistent densities throughout the entire kVp range.

1. Select 120 kVp, keeping all other settings the same from the previous procedure. Add additional phantom material to simulate a large patient (i.e., 12 inches total acrylic or water).
2. Take an exposure with film loaded in the cassette. Make sure exposure time is greater than 30 ms (lower the mA setting if it is not). Develop the film and measure the density. If it is not 1.2 D (nominal), adjust the kVp compensation potentiometer (high kVp adjustment) until that density is achieved.
3. Select 60 kVp, 200 mA large focus, and reduce the phantom thickness to simulate a small patient (i.e., 6 inches of acrylic/water). Take an exposure, develop the film, and measure density. Adjust the kVp compensation potentiometer (low kVp adjustment) for a density of 1.2 D. Recheck 120 kVp since these adjustments may interact

Short Time Compensation (For PMT Only)

The response of the PMT is not linear at very short exposure times. Consequently, phototimers must have additional circuitry to compensate for this linearity problem. The procedure for adjusting the short time compensation circuit is described below.

1. Select 80 kVp, 8 inches of acrylic, and the zero density setting.
2. Select an mA setting that will achieve an exposure time that is less than 10 ms.
3. Take an exposure and develop film. Adjust the short time compensation potentiometer for 1.2 D.

Finally, the service engineer must check to see if the density compensation circuit (+/– D) is working properly. Check the manufacturer's specifications for density increments. For example, a + 1.25 setting on control should produce a film that is 25 percent darker than the "0" setting. Check this for all density compensation selections.

FINAL TESTING

Now that all the devices in the x-ray room have been calibrated and adjusted to specifications, the room should be ready for use. Before the room can be "turned over," however, two important tasks must be performed. Because of the numerous alignment and adjustment procedures that were performed by the service engineer(s) during the installation, it is possible that some items could have been overlooked. A ***final room checkout*** should therefore be performed to ensure that all the equipment is operating as specified by the manufacturer. When the final testing has been completed, *the room should be thoroughly cleaned and polished.*

The final check should be performed by a senior service engineer who was not one of the actual installers. This is done to eliminate any bias that may be present when testing the room. All items in the room should be tested during the final room checkout. Below is a checklist to be used for the checkout procedure. The service engineer can use this same checklist as a final checkout after a preventive maintenance has been performed. This check should be performed along with the compliance testing that is required by law.

INSTALLATION/PM CHECKLIST

The Tube Support

1. Inspect wiring and electrical connections.
2. Inspect cable drapes.
3. Inspect tube support movement.
4. Check that all locks are adjusted properly.
5. Inspect counterweight assembly.
6. Inspect the tube mounting hardware.
7. Check tube alignment to receptors.
8. Check all lamps and indicators.
9. Check SID indicator accuracy.
10. Check that all covers are in place and fastened.

The Table

1. Inspect tabletop movement.
2. Inspect vertical drive.
3. Inspect operation of locks.
4. Inspect bucky and cassette tray.

5. Check any safety interlocks.
6. Check all lamps and indicators.

The Wall Receptor

1. Inspect vertical movement.
2. Check lock adjustment.
3. Inspect counterweight assembly.
4. Inspect bucky and cassette tray.

The Generator/Control

1. Check that all electrical connections are secure.
2. Inspect all indicators and lamps.
3. Check that all modes of operation are functional.
4. Inspect all switches, buttons, and knobs.

The High Voltage Transformer Unit

1. Inspect the high voltage cables and connections.
2. Check all the generator interface connections.
3. Check oil level in transformer.
4. Check that all covers are in place and fastened securely.

The X-ray Tube

1. Check that the high voltage cable ends are clean, sufficiently greased, and are securely tightened.
2. Check rotor bearings (sound and speed).
3. Check x-ray calibration (noninvasively)–record calibration results on a calibration form.
4. Take a test film using a resolution phantom such as a pie mesh or line pair phantom to check tube resolution.
5. Check the focal spot size of the tube.

The Collimator

1. Test lamp operation.
2. Verify SID indicator accuracy.

3. Check light field accuracy.
4. Check actual x-ray field size with field size indicators.
5. Check the operation of the automatic system.

The AEC

1. Test AEC operation with a flat field phantom.
2. Check for the proper density (i.e., 1.2 D).
3. Test for exposure repeatability.
4. Test all three chambers for correct balancing.
5. Test back-up timer function.

Finally, the service engineer should inspect all equipment surfaces for marks, dents, or scratches that could have been caused during the installation. If necessary, the engineer should "touch up" any surface defects. All surfaces should be cleaned thoroughly to remove grease and fingerprints. A good amount of time should be spent on cleaning the equipment so that it really shines. Since the customer has most likely paid a lot of money for the new x-ray equipment, it should be in "showroom" condition at the completion of the installation.

In addition, the service engineer should make a list of any outstanding items regarding the room that need to be addressed. These items can be corrected during the follow-up phase of the installation (see below).

USER INSERVICING

Before the room can be turned over for use, the hospital staff must be taught how to safely operate the equipment. The demonstration of equipment operation is referred to as ***user inservicing.*** Inservicing of the medical staff is a very important part of the installation process. In fact, the success or failure of a newly installed room is often dependent on how comfortable the staff is with the operation the new equipment.

Many service organizations fail to realize the importance of user inservicing. It is common to hear of situations where a new room was installed at a hospital, but the technicians much preferred to use another "older" room instead. When asked why this was so, they would respond with a long list of complaints about the equipment–most of which were operational problems. The staff perceived the problems as equipment related and, as a result, had a negative view of the equipment (and the equipment dealer, as well). A comment often heard is that the dealer made the sale and then abandoned the customer.

User inservicing should be performed by an experienced member of the equipment dealer, usually designated as an ***applications specialist.*** A senior x-ray service engineer can perform this task if knowledgeable in radiographic positioning and filming techniques. The equipment dealer, together with the chief x-ray technician at the facility, should make a schedule for inservice training after the final room checkout has been completed. Inservices should be offered at different times on all shifts to ensure that everyone can attend a session. An attendance list should be used at each inservice to ensure that all staff members have been inserviced.

Proper inservicing must always include a review of the ***operation manuals*** so that the technicians will know exactly where to look up in information if they are experiencing difficulties with the equipment. The representative who is giving the inservice should then provide a ***hands-on demonstration,*** showing the basic operation of the equipment, safe handling, and any special features that are available on the unit. Finally, the applications person should be present when the equipment is first used on actual patients to ensure that the equipment operates to the customer's satisfaction and also to provide assistance, if needed.

FOLLOW-UP

After the room has been in use for one month, a follow-up visit is needed to check that the room has been operating properly and, to see if the customer has encountered any problems with the equipment. Also, there is always a "break-in" period with new equipment where certain hardware will loosen from use, or calibration adjustments can drift slightly. Usually, the x-ray tube calibration must also be checked and adjusted at this time.

It is at this time that the hospital staff will provide the service engineer with the list of additional items that they would like addressed. It is very important to examine this list carefully before taking any service action. Many of the items on the list may be related to one problem area, or the list could contain problems that were operator related and should be addressed by the applications specialist. The list could also indicate symptoms of a serious failure with the equipment.

After the list is carefully scrutinized, corrective action can be taken. As each item is resolved, it should be checked off from the list. This is done so that when the service is completed, the service engineer can demonstrate the corrective action that was taken for each point on the list.

Chapter VII

PREVENTIVE MAINTENANCE

BACKGROUND

Equipment, in general, will tend to degrade in performance when used regularly over a period of time. Mechanical parts will wear out due to friction and the electrical characteristics of electronic components will change over time as a result of being exposed to heat and current fluctuations. Eventually, certain components of the x-ray system will reach their useful performance life and then just cease functioning. Furthermore, as the components of an x-ray system slowly age, the quality of the images that are produced will gradually begin to degrade.

Radiographic equipment has a predictable life expectancy that is determined by the equipment manufacturer and is termed *useful life.* During its useful life, the equipment is expected to perform to the specifications stated in the equipment manuals. In order for the equipment to maintain this optimum performance level, periodic inspections and adjustments must be performed by a trained x-ray service engineer. In fact, if every component of an x-ray system is inspected on a routine basis, the equipment will perform reliably for well beyond its predicted useful life.

The periodic inspection and adjustment of equipment is called ***Preventive Maintenance*** or ***PM,*** for short. A PM on a radiographic room should be performed by an x-ray service engineer at a frequency of at least once a year for most radiographic equipment (every 6 months for equipment that is used excessively). As a rule, a complete PM on a general radiographic room takes one full day (8.0 hours) for one engineer to complete.

When a particular radiographic room is due for a PM, the engineer will usually call to schedule a convenient time to perform the service. This fact–that a preventive maintenance is ***scheduled downtime***–is one of the main reasons for performing a PM and must be emphasized to the customer. Scheduled downtime means that the radiologist and staff at the facility know in advance that the room will not be available for use and have made arrangements to x-ray their patients in another room or at another site. Though it is sometimes difficult to schedule this downtime, especially at a busy facility, scheduled downtime is a much better alternative than the emergency downtime caused by equipment failure.

When equipment fails during regular working hours, a serious and usually stressful situation results for both the hospital staff and the service engineer. In this situation, the patients at the office will have to be sent home, and all remaining appointments must be canceled and then rescheduled for a future time. This creates problems for both the staff, who now can't "x-ray" patients but are nonetheless being paid to do so, as well as for the patients who have to rearrange their schedules and drive sometimes very long distances back to the facility. *Performing a PM on a regular basis effectively reduces emergency downtime.*

A final and equally important point relates to ***patient safety.*** It is inevitable that some of the mechanical parts of x-ray system will eventually loosen with use. It is not uncommon to hear of situations where a piece of hardware has fallen from the x-ray tube assembly or from the table and injured a patient or staff member. Invariably, this situation occurs on equipment that has not been routinely inspected. Also, a failure in a device with a motor drive system could result in serious injury to the staff or patient. For safety reasons, such devices must be inspected regularly. One way the engineer can put this concept into perspective is to ask the question: "Would I feel comfortable if a member of my own family were being x-rayed in this room?"

Overview

During the PM, the service engineer will closely inspect each device in the room, checking for any hardware that has loosened over time and for parts that show signs of wear. All cables and their connections to each device will be inspected, cleaned, and tightened to eliminate the chance of intermittent failures due to faulty connections. The gears of motor assemblies will be inspected and lubricated, and all belts and chains adjusted. All the indicator lamps and control knobs will be checked to see if operating properly and will be replaced if necessary.

The x-ray tube will be calibrated during the PM. The reason for the calibration is partly due to electrical instabilities in the x-ray generator, but mostly it is necessary because of the "filament drift" that occurs with all x-ray tubes.

Each time a filament wire is heated, minute particles of tungsten become vaporized and are then deposited on the inside of the glass insert of the x-ray tube. This evaporation of the filament causes an increase in resistance of the filament wire which, in turn, lowers its heating value. During the x-ray tube calibration, the filament supply voltage will be increased to compensate for the filament loss.

The collimator will be tested during the PM and the AEC is checked. When all the adjustments are completed the engineer must perform all compliance testing required by state and federal regulations. Since guidelines vary from state to state, the engineer must obtain the appropriate documentation for medical diagnostic x-ray equipment. This information is available through government agencies, such as the ***Bureau of Radiological Health*** (or ***BRH***).

Finally, after all the adjustments are completed, *all equipment surfaces should be cleaned.* This is very important because it is the first thing that the staff notices following a PM. The fact is that the radiologist or the institution has just paid a significant sum of money for the PM and expects to see some physical indication of the work that was performed. They cannot actually see most of the adjustments and calibrations that were performed and, consequently, need some visual evidence that great attention was paid to their equipment and that it was treated with care.

For instance, the same principle applies when an automobile is taken to the dealer for the 30,000 mile service checkup. When the hood is raised it is a psychological plus for the dealer if the owner can see obvious signs that someone was actually working in this area. The spark plugs and filters should appear new or at least clean. The reader will also agree that greasy fingerprints left on the hood of the car would be absolutely unacceptable.

Before beginning the PM, the service engineer should consult the radiologist and/or chief technician to find out if he or she has been experiencing any problems with the room. This is done so that the engineer can concentrate on those particular areas and directly address problems encountered there.

Many times a device may be operating at an acceptable performance level, but the hospital staff would like it adjusted to their own preferences. Without first consulting the staff, the engineer would not normally make any adjustments to the device and, consequently, will have to return to address that particular complaint. This practice wastes time for both the engineer and staff and often reflects badly on the individual service engineer.

If the service engineer encounters any damaged or faulty devices, he or she will have to assess the situation and decide whether to repair the device at that time or perform the repair at a later date. If a replacement part must be ordered, the engineer will, of course, have to return with the new part at a later date. Minor repairs, those taking less than an hour, should be performed during the PM.

Many service engineers feel that since so many other duties must be performed during the PM, it is best to make notes of the repairs that are needed and schedule another day in the room for these repairs. This approach is fine if the staff at the facility is willing to cooperate. In addition, the engineer should always get the approval of the customer before ordering any expensive parts.

The PM checklist is the same one that was used for the final room checkout at the end of the installation chapter. For convenience it is reproduced at the end of this chapter.

PM PROCEDURE (DETAILED DESCRIPTION)

Mechanical Inspection

It is best to have an orderly routine when performing the PM. A good PM routine should start with the ***cleaning and inspection*** of all the mechanical assemblies and all the electric connections in the radiographic room. This includes the tightening or replacement of any hardware that has loosened or is missing, as well as the removal of dirt, dust, and other foreign materials from the mechanical assemblies. All bearing and bearing tracks should also be cleaned and lubricated at this time.

The service engineer should make a general inspection of all switches, locks, solenoids, and relay contacts, and correct any problems that he or she encounters. All the connections on the x-ray generator should be inspected and tightened since a loose connection could affect the calibration. Next, the belts and/or chains in the motor drive assemblies should be inspected and adjusted, if necessary. The counterbalancing systems and motor assemblies should be checked and lubricated at this time.

Mechanical inspections are always performed *first* for several reasons. First of all, it is important to have good, clean connections before attempting any adjustments. Many times a loose connection will affect the reliability of the adjustment and could also cause intermittent failures. If the engineer notices an intermittent problem during the calibration and adjustment phase, he or she will then have to stop the procedure to find the bad connection. This results in wasted time because once the problem is corrected, the engineer must completely redo the calibration.

Another reason for performing mechanical checks first is to ensure that all gears, railings, tracks, and bearings are sufficiently tightened before the alignment checks are made. As described in the last chapter, bearing adjustments will affect the alignment of the devices.

Also, a lot of dust is created when the ceiling rails and upper sections of the tube support are cleaned. These overhead areas cannot be reached without a ladder and, consequently, only get cleaned and inspected by the service engineer. The tube support should be "PM'ed" first so that the dust, can settle down before cleaning any other equipment. *A dust mask (or surgical mask) should be worn during the cleaning phase if the dust is excessive.*

Once the mechanical and electrical connections of the tube support have been inspected and adjusted, the engineer can perform similar checks on the radiographic table and wall receptor. When inspecting the table, the engineer should be sure to check the safety interlock switches that may have become loosened or damaged since the last PM was performed.

High Voltage Cable Maintenance

The service engineer should carefully examine the high voltage cables to see if there are any visible cuts in the rubber insulating cover.

Also, the cable ends should be cleaned (discharge first!) and the individual pins of the federal connector should be spread apart if needed. Pin spreaders are available from some equipment manufacturers, but a pocket knife can be used as well.

In either case, caution must be exercised when spreading the pins because they are fragile and can easily break off. Since high voltage cables are no longer repaired in the field, a cable with a damaged pin will have to be replaced. This mistake will result in an unexpected cost to the customer (high voltage cables are very expensive). In addition, the room could be "down" for several days while two engineers perform the often difficult task of pulling the high voltage cable through conduit that is now filled with many other cables.

In an effort to eliminate the problem of maintaining good pin contact in the high voltage well, some x-ray tube manufacturers are using a newly-designed receptacle that contains a spring in each of the pin sockets of the x-ray tube. This design ensures good pin contact and eliminates the need to spread the pins of the high voltage cable. In addition, a few cable manufacturers are now selling high voltage cables that have replaceable pins. This new design eliminates the aggravation caused by broken pins.

When the cables have been inspected and cleaned, new ***vapor proofing compound*** must be applied to the federal connectors that are inserted in the x-ray tube. Vapor proofing compound, also known as dielectric compound or tube grease, is used to insulate the individual pins of the connector from ground. When properly applied, the tube grease will completely surround the connector when it is plugged into the tube well. The compound forces the air (a good conductor for high voltage) out of the well, thus eliminating any chance of arcing.

The compound should be applied with an applicator (e.g., wood tongue depressor) to the end of each federal connector. This is done to eliminate any contamination (i.e., salt or oils) that could be introduced from the hands of the service engineer. Latex gloves, available at most medical facilities, can be worn while applying the compound to ensure that no foreign materials are introduced to the cable ends.

Enough compound should be used to form a mound on the end that completely covers the pins (Figure 27). The compound should then be applied evenly to 1/2 of the phenolic insulator portion of the federal connector. After the tube wells have been thoroughly cleaned, the cable end can be inserted into the well by aligning the "key" of the

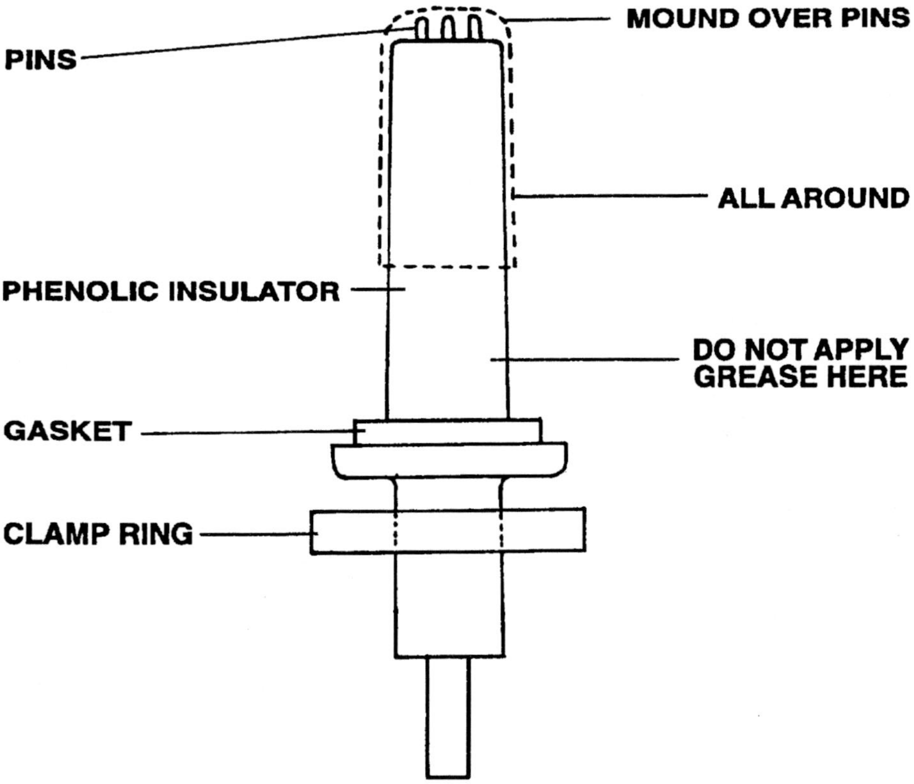

Figure 27. Application of vapor proofing compound. Correct application of the compound assures that air gaps will not form in the x-ray tube receptacle.

cable with the "slot" in the tube well, and then pressed firmly into place.[21]

The retaining ring nut should be threaded on until the rubber o-ring is tightly compressed. The rubber o-ring acts like a protective seal that keeps moisture out of the tube well and also helps prevent the grease from drying out. If Allen screws are provided on the retaining nut, these are tightened at this time. After the tube grease has been warmed during the x-ray tube calibration, the rings should be retightened a final time to ensure that it is completely seated and all the air is removed from the well.

21. The engineer should check that the grounding spring is in place! The grounding spring is the circular metal washer that is slightly bent to act like a spring. This spring must be installed to ensure grounding integrity and thus avoid high voltage problems.

The engineer should perform the same maintenance on the federal connectors at the high voltage transformer. Dielectric compound, however, should not be used in these wells. Instead, ***transformer oil*** is used because it is a much better insulator than the dielectric compound and does not degrade as quickly. The wells in the transformer should be filled to a depth of approximately 1/2″. When the cable is plugged in, a small amount of oil should ooze out, confirming that the cable is sufficiently insulated.

Ideally, transformer oil would be used in the wells of the x-ray tube because it is a better insulator. However, since the oil could leak out whenever the tube was rotated from the upright position, the more viscous dielectric compound is used. Although the vapor-proofing compound successfully addresses the leakage problem, it is only a mediocre substitute for the oil.

Vapor-proofing compound will break down chemically and will dry out when exposed to heat. As the compound dries, air gaps will form in the high voltage well providing a path for arcing. It, therefore, must be replaced, without fail, on every PM. Transformer oil does not have to be replaced nearly as often, but should nonetheless be inspected periodically.[22]

Some manufacturers of high voltage cables provide a molded rubber insulating disc that fits directly onto the cable ends. The advantage to using the disc is that it eliminates the need for vapor proofing compound. Instead, the discs are lightly coated with a small amount of silicone oil and placed on the end of the high voltage cables. The cables are inserted in the wells and are tightened to compress the disc.

The discs are a good alternative to the grease and definitely less messy. However, each disc can only be used one time! If a cable is removed from the well, for any reason, a new disc must be installed. The reason for replacing the disc relates to the fact that the disc becomes permanently distorted when compressed and will not seat properly if used a second time.

When the service engineer has completed the mechanical and electrical inspections, the system can be "powered-up" so that the lamps and switches can be checked. In addition, all other generator functions

---

22. Transformer oil is much more durable than the dielectric grease but will break down over time with exposure to heat. When the oil degrades, it will become discolored, changing from a clear off-white color to a dark yellow-orange. The oil should be changed when the discoloration is readily apparent.

such as tube safety circuits and memory function can also be tested at this time.

This is a convenient stopping point in the PM procedure to take a break. With all the mechanical inspections performed, the PM is halfway completed. The second part of the PM will continue first with the calibration adjustments, then followed by alignment checks, image quality checks, repairs, compliance testing, and finally, equipment cleaning.

Electronic Adjustments

The second half of the PM should begin with the calibration of the x-ray tube. For this calibration, the service engineer can use a noninvasive kVp device for kVp measurements and a digital mAs meter for invasive mA/mAs measurements. If the generator provides actual kVp and mA feedback test points, they can be used since their accuracy should have been verified at installation.

The x-ray tube should be warmed up before proceeding with the calibration by using the procedure outlined in Chapter VI (see tube warm-up). The calibration should begin on the large focus first, followed by the small focus. This is the accepted sequence used for calibration since the small filament is much more fragile than the large filament and is more easily damaged. Hitting the small filament of a cold tube, with even a moderate exposure technique, could stress the filament enough to cause permanent damage. Because the large filament is much more durable, it is always calibrated first.

Care should be taken so as not to overheat the x-ray tube. If a 15 to 20 second interval between exposures is used, the tube should remain relatively cool and should never approach its maximum heat capacity. Even so, the tube housing should be periodically checked (by touch) during the calibration to make certain that it is not overheating. If the tube feels hot to the touch, the calibration should cease for 15 minutes to a half hour. As explained in the troubleshooting discussion in Chapter IX, heat contributes to a majority of x-ray tube failures.

During the calibration the engineer should listen to the rotor bearings and note any excessive noise. The boost/run time was setup at installation and should be correct, but it must be checked by either measuring the boost time and comparing it with the equipment manufacturer's specifications, or the actual rotation should be checked with a vibrating reed tachometer. Usually the engineer will notice obvious sound changes between successive boosts that would indicate an incorrect adjustment.

When the calibration is completed, the calibration results should be recorded and checked to see that they fall within the manufacturer's guidelines. Any values that are out of range must be recalibrated.

Following the x-ray tube calibration, the collimator should be tested. The engineer should follow the collimator checklist found in Chapter VI. The tube/receptor alignment should be checked at this time, and the centering switches adjusted if necessary.

The AEC should be checked for accuracy and repeatability. However; if the customer is happy with the current film densities, *DO NOT ADJUST THE AEC CONTROL!* Whenever a conscientious service engineer decides, with perfectly good intentions, to fine tune an AEC system that is already performing satisfactorily, he or she will invariably have to return the next day to "change it back to the way it was before!"

In the last step of the PM, the service engineer will clean every equipment surface. All equipment cabinets and especially the radiographic table top should be free of grease and any hand prints acquired during the PM. The engineer should then take the time to remove the *adhesive tape* (x-ray technicians use adhesive tape for a variety of creative purposes!) from the equipment surfaces and clean all tacky surfaces caused by the residual glue from the tape. A multipurpose cleaner can be used on most equipment surfaces to remove dirt and grease. If there are any major scratches on the equipment, they should be covered with touch-up paint. Also, the floor must be checked for oil and other lubricants that may have spilled during the PM. These spills can be cleaned with alcohol or other approved solvent.

INSTALLATION/PM CHECKLIST

The Tube Support

1. Inspect wiring and electrical connections.
2. Inspect cable drapes.
3. Inspect tube support movement.
4. Check that all locks are adjusted properly.
5. Inspect counterweight assembly.
6. Inspect the tube mounting hardware.
7. Check tube alignment to receptors.
8. Check all lamps and indicators.
9. Check SID indicator accuracy.
10. Check that all covers are in place and fastened.

The Table

1. Inspect tabletop movement.
2. Inspect vertical drive.
3. Inspect operation of locks.
4. Inspect bucky and cassette tray.
5. Check any safety interlocks.
6. Check all lamps and indicators.

The Wall Receptor

1. Inspect vertical movement.
2. Check lock adjustment.
3. Inspect counterweight assembly.
4. Inspect bucky and cassette tray.

The Generator/Control

1. Check that all electrical connections are secure.
2. Inspect all indicators and lamps.
3. Check that all modes of operation are functional.
4. Inspect all switches, buttons, and knobs.

The High Voltage Transformer Unit

1. Inspect the high voltage cables and connections.
2. Check all the generator interface connections.
3. Check oil level in transformer.
4. Check that all covers are in place and fastened securely.

The X-Ray Tube

1. Check that the high voltage cable ends are clean, sufficiently greased, and are securely tightened.
2. Check rotor bearings (sound and speed).
3. Check x-ray calibration (noninvasively)–record calibration results on a calibration form.
4. Take a test film using a resolution phantom such as a pie mesh or line pair phantom to measure tube resolution.
5. Check the focal spot size of the tube.

The Collimator

1. Test lamp operation.
2. Verify SID indicator accuracy.
3. Check light field accuracy.
4. Check actual x-ray field size with field size indicators.
5. Check the operation of the automatic system.

The AEC

1. Test AEC operation with a flat field phantom.
2. Check for the proper density (i.e., 1.2 D).
3. Test for exposure repeatability.
4. Test all three chambers for correct balancing.
5. Test back-up timer function.

Chapter VIII

IMAGE QUALITY

WHAT IS IMAGE QUALITY?

In the field of x-ray servicing, each alignment and calibration procedure is performed with one goal in mind: ***Image Quality.*** After all, the purpose of x-raying a patient is to obtain an image of a specific internal structure that can be later used for medical diagnoses. The goal of the x-ray service engineer is to "tweek" (fine tune) the x-ray system so that the best possible image (for that system) can be produced. To intelligently discuss image quality, however we must first define what is meant by a "good image."

To begin with, a good image should have sufficient ***optical density*** so that the internal structures can be clearly seen on the imaging medium (i.e., film). Optical density refers to the overall "blackening" of the film and is the most important factor of a quality image. The image must be dark enough so that the all structures of the patient can be clearly seen when viewed on an ***illuminator.***

Optical density is precisely measured with a ***densitometer,*** which can detect very small changes in film density. The densitometer measures film density by passing a light of a certain intensity through a small area of the film and then compares it with the intensity of light that emerges from the film. The film density is equal to the common logarithm of the ratio of the two measured intensities. Thus $D = \log (I_o/I_t)$, where I_o is the original calibrated light intensity and I_t is the intensity of the light transmitted through the film.

The human eye can only distinguish density differences in a very limited range (up to about 2.5D). In fact, at an optical density above 2.5D, the film will be so dark that no specific structures in the image can be distinguished. In addition, at extremely high (or low) densities, the ***contrast range*** of the film is significantly reduced (discussed later). Normal film densities range from approximately 0.5D to 1.5D.

An image should also have sufficient ***contrast*** so that each type of tissue (bone, cartilage, soft tissue, etc.) can easily be discerned. Image contrast refers to the difference in optical density between two adjacent structures on the image. A greater density difference between these adjacent structures will result in higher image contrast. A high contrast image will appear black and white with very few shades of gray. A low contrast image will gradually change from black to white, having many shades of gray in-between.

The optimum contrast level for an image will vary with the type of anatomy being filmed. If the radiologist is interested in viewing bone tissue, an image with high contrast is preferred. When soft tissue is being studied, a lower contrast image is needed to distinguish the various individual structures.

And finally, a good image should have a sufficient level of ***detail*** so that the borders of each structure appear sharp and focused. Image detail refers to how quickly the change in density occurs between two adjacent regions. This directly relates to how sharp the individual structures will appear. A very abrupt change in density between adjacent areas on the image will produce an image with clearly defined sharp edges. When discussing image quality, an image that has structures with clearly defined, sharp edges is said to have ***good detail.*** Other terms used interchangeably when referring to image detail are ***image definition, image resolution,*** and ***image sharpness.***

In addition to density, contrast, and detail, a good image should precisely represent the actual size and shape of the patient's anatomy and should not "distort" the internal structures of the patient. ***Image distortion*** occurs when the patient is not properly positioned in the x-ray field. If the final image is distorted, a misdiagnosis could result.

Finally, a good image should have minimum background "noise." ***Radiographic noise*** has a "grainy" or "salt and pepper" appearance on the film. The noise is a result of several factors discussed later in this chapter. Radiographic noise only reduces image contrast and will, therefore, greatly deter from image quality.

Each device in the x-ray system has some effect on image quality and will be discussed in detail in the following pages. In the previous two chapters we discussed the procedures used to align and calibrate each of the various radiographic devices. This chapter will demonstrate the actual effects the adjustments have on image quality, as well as all other factors that affect image quality. This information will aid the service engineer in improving the images produced from an x-ray system. However, before discussing the effects that each device has on image quality, a brief overview of the image formation process is needed.

IMAGE FORMATION

When discussing image formation, we must consider all aspects of radiography, from the production of x-rays to film processing. Of all the factors considered, it will be shown that the most important factor in the image formation process is ***the quality of the x-ray beam*** itself. In order to produce a good image, the beam emanating from the x-ray tube must be of sufficient energy (i.e., quality) to pass through the patient to the imaging medium. Once the proper "beam quality" is achieved, each device in the imaging pathway will only modify the x-ray beam in some way to produce the final image.

In order to record a diagnostic image of a patient's internal structures, an x-ray beam of a sufficiently high energy level is needed. When this beam strikes the patient the x-ray photons will interact with the tissues in the body in three different ways. The highest energy photons will completely ***penetrate*** the patient, passing directly through to the image receptor. The remaining photons of lower energy levels will be ***attenuated*** in two ways. When striking the tissue, the photons will either be ***absorbed*** directly into tissue or they will ***scatter*** (i.e., travel in a direction other than that of the primary beam).

It is the differential absorption of the x-ray beam that essentially creates the image. Thus, the discussion of image formation will concentrate in this area only. Scatter radiation only hinders the image formation process and will be discussed in detail later in this chapter.

The internal structures of the patient will absorb x-ray photons differentially, depending on the density of the particular tissue. The variation in absorption of the beam will result in a beam exiting from the patient that consists of radiation of varying intensities. Specifically, the energy level of the x-ray photons in the ***exit beam*** will vary per cross-sectional area of the patient. This variation of intensities, known as ***sub-***

ject contrast, will form the actual image on the film. Subject contrast will, in turn, determine the amount of image contrast.

The exit beam that emerges from the patient passes through the ***intensifying screens*** of the image receptor to the film. These screens, made of a luminescent material, will convert the x-ray photons to light photons that will expose the film. The film consists of a flexible ***base,*** usually made of plastic, which has an adhesive coating to which the ***emulsion*** will adhere. The emulsion consists of sensitive silver bromide crystals suspended in a gelatinous material.

When the beam strikes the intensifying screens, the x-ray photons are converted to light energy. The light energy that hits the emulsion of the film causes the silver bromide crystals to undergo a chemical change resulting in the depositing of neutral silver on the film. The amount of silver that is deposited at each point on the film is directly related to the amount of light that strikes the crystals at that point. The clustering of neutral silver deposits in the emulsion forms an invisible image that is termed the ***latent image.*** The crystals that are not exposed to light will remain physically unaltered.

The exposed film with the latent image must undergo further processing in order to make the latent image visible to the human eye. The film is "processed" in a solution (i.e., ***developer***) that contains more silver ions. The silver ions in the developer react with the neutral silver to form deposits of black metallic silver. The developer has no effect on the unaltered silver bromide crystals. The black silver grains, deposited in the gelatin of the emulsion, form the visible image.

When the film has been fully developed, it must then be placed in another chemical, the ***fixer,*** which clears the film of all unaffected silver bromide crystals. If these unaltered crystals were to remain on the film, they would further react with light and, consequently, fog the film. The fixer also hardens the film and makes the image permanent.

After the film is washed and then dried, the image formation process is complete and the film is ready for viewing. Fully processed film, if stored properly, will remain in diagnostic condition for many years.

BEAM QUALITY

As stated earlier, the quality of the x-ray beam is fundamental to obtaining good image quality. Beam quality describes the overall energy level of radiation in the x-ray beam which will indicate the ***penetrating ability*** of the beam for a given kVp. If the beam consists predominantly of high energy photons, it is considered to be a "hard" beam. A hard beam will have good penetrating ability since more of the high energy photons in the beam will pass through the patient unimpeded. Furthermore, if more photons are passing through the patient, fewer photons will be absorbed by the patient.

If the beam consists of lower energy x-ray photons, it is referred to as "soft" radiation. Soft radiation has a lower penetrating ability and is mostly absorbed by the patient. Because the low energy photons are more easily absorbed in tissue, a soft beam will increase subject contrast and for this reason is desirable in special applications of radiography such as mammography, where the visualization of soft tissue is desired.

The three major factors that affect the quality of the x-ray beam are the selected ***kVp*** (and the ***kVp waveform***), the material used in the anode of the x-ray tube (or ***anode construction***), and the amount of ***filtration*** in the x-ray beam. The effects of kVp and anode construction are discussed in the following sections. Our discussion now concentrates on filtration and its relation to beam quality measurements.

FILTRATION

Filtration is the process of removing unwanted low energy radiation from the x-ray beam. The x-ray beam is filtered by placing thin sheets of homogenous metal, usually aluminum (AL_{1100}), in the path of the beam at a point as close to the x-ray tube as possible, usually at the collimator. Placing the filter this close to the tube effectively blurs out any imperfections (i.e., scratches) in the filter.

The aluminum "filters" the beam by removing the low energy x-ray photons from the beam. These low energy photons do not contribute to the image in general radiography and, more importantly, increase the dose delivered to the patient. *By the use of filtration, the penetrating ability of the beam can be increased and, as a result, the patient will absorb much less radiation.*

Moreover, as more filtration is added in the path of the beam, the penetrating ability of the beam will correspondingly increase. Excessive filtration results in an image with extremely high contrast which is not desirable for viewing soft tissue. Again, the most significant drawback with using too little filtration is that the patient absorbed dose will increase.

MEASUREMENT OF BEAM QUALITY

Beam quality is defined in terms of the amount of filtration in the x-ray beam. The measurement of beam quality is called the ***Half Value Layer*** or ***(HVL).*** The half value layer is defined as *the amount o aluminum required to reduce the radiation output of the tube to one half the original value at given kVp.* An important point to remember regarding HVL is that beam quality is very dependent on the kVp used (see Technique Selection below) and accordingly, so is the HVL measurement. Therefore, the kVp setting must always be recorded with the HVL result.

The half value layer is expressed numerically in terms of the "equivalent amount of aluminum" (in millimeters). A soft x-ray beam can be significantly reduced by a small amount of aluminum and, therefore, has a low HVL. Similarly, a hard beam will require a large amount of aluminum to reduce the radiation to one-half and has a high HVL. Ideally, what is needed for a good image is a beam with a minimum HVL to ensure patient safety but one that contains photons at many different higher energy levels so that some photons will penetrate the tissue while other photons will be absorbed. A ***polyenergetic*** beam will produce the best image contrast.

Guidelines for the minimum values for HVL for each selected kVp are set by the federal government. These guidelines are established to ensure patient safety and all x-ray units must be in compliance with these standards. A list of the half value layer requirements can be found in the Appendix. The service engineer must use the half value layer measurement to determine the minimum amount of filtration that must be added or removed from the beam to meet these guidelines.

FACTORS THAT INFLUENCE IMAGE QUALITY

Figure 28 shows the typical orientation of the x-ray imaging chain for a radiographic room. Each of the highlighted items (and their geometrical relationship to one another) produces a significant effect on the

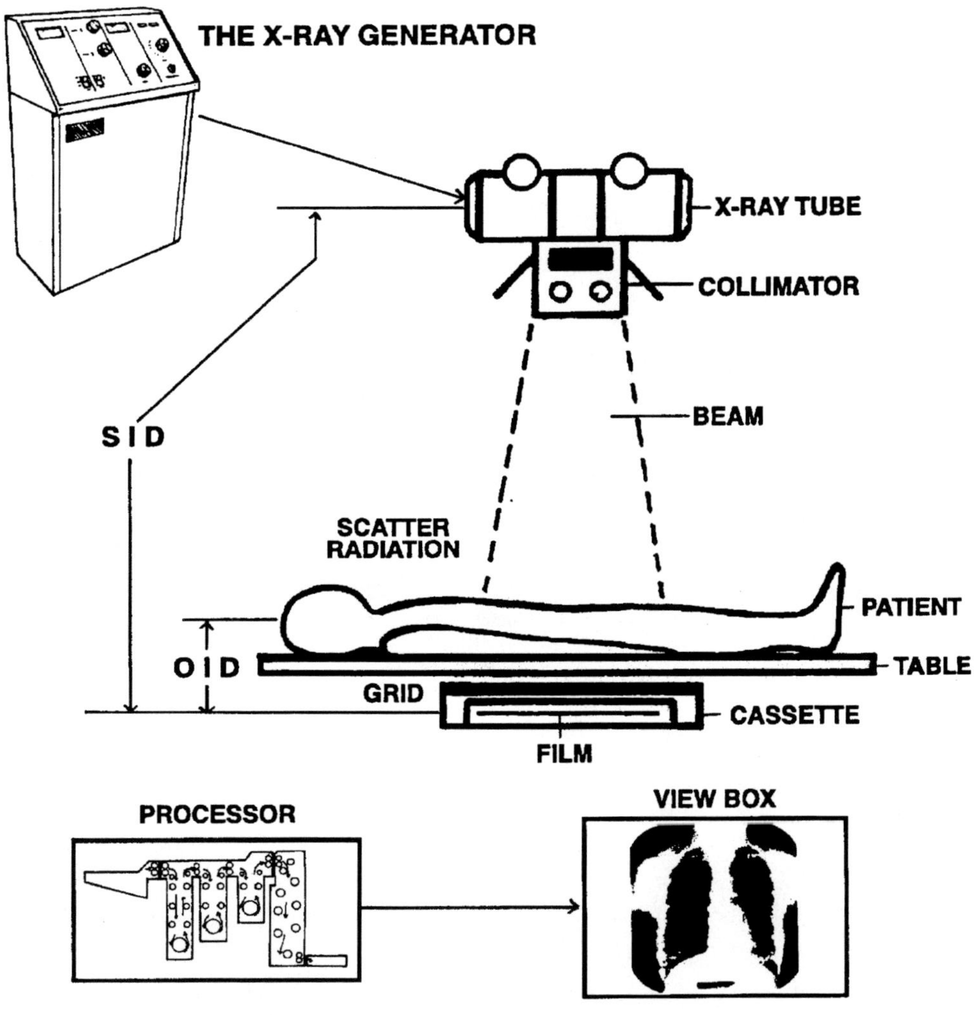

Figure 28. Factors that influence image quality. Each of these components has a significant effect on the quality if the final image.

final image. After the image is formed on the x-ray film and the film has been processed, the last factor affecting image quality is the viewing condition.

The X-ray Generator (Technique Selection)

Technique selection refers to the technique factors of ***kVp, mA,*** and ***time*** that are selectable at the x-ray control. These factors greatly influence the quality and quantity of radiation delivered to the patient per exposure and, therefore, directly affect image quality.

The selection of kilovoltage has two effects on image quality: it directly affects the ***beam quality*** which will greatly influence image contrast, and it also affects the overall film density. As stated above, beam quality refers to the overall energy level of the radiation present at a given kVp. At very low kVp settings, the beam consists predominately of soft radiation which will be absorbed by the patient and, therefore, will not contribute to the x-ray image. At high kVp settings, the beam consists of high energy x-ray photons which have greater penetrating ability. This hard beam will effectively reduce the dose to the patient but at a cost: a ***loss in image contrast.***

As a rule, *increasing the kVp will reduce image contrast.* The loss in contrast is a result of: (1) more x-ray photons reaching the image receptor (because of the higher penetrating ability) thus darkening the film overall and (2) increased amount of scatter radiation caused by the high energy photons. Scatter radiation, as stated above, only reduces image quality.

A radiograph taken at 80 kVp that produces a normal film density will have noticeably more contrast than one taken at 100 kVp for that film density, all other factors held constant. This effect on image contrast applies mainly to large changes in the kVp setting. The human eye cannot detect a change in contrast when small changes in the kVp settings are made. Specifically, when the exposure rate (film density) is held constant, as in the case when AEC is used, a change in +/− 5 kVp is not discernible to the unaided eye.

Second, an increase in kVp will cause more photons to reach the film, and thus a greater ***film density.*** The increase in the ***quantity*** of radiation is, however, *not proportional* with the increase in kVp. For example, doubling the kVp from 50 kVp to 100 kVp will not double the film density–all other factors held constant.

Finally, the quality of the x-ray beam is affected by the type of ***kVp waveform.*** As stated in Chapter I, the kVp waveform, when viewed invasively, is drastically different for single-phase generators, three-phase, and high frequency generators (see Figures 4, 5, 6). At 80 kVp, the x-ray beam will be much "harder" for three-phase than for single phase because of the reduced ripple factor and higher efficiency of three-phase power. Consequently, a single phase generator will produce more image contrast than a three-phase or high frequency generator at that same kVp and film density.

The technique selection of ***mA*** and ***time*** also affects image quality, but in a way not related to image contrast. When referring to image quality, mA and time, or the product of mA and time (***mA*** x ***seconds*** or ***mAs***), directly affects ***film density*** and ***image sharpness.*** If the kVp is held constant, increasing a mA setting or the time setting will *proportionally* increase the overall density of the image.

The mA and time factors also indirectly affect how sharp the image will be. The sharpness of the image can be affected by three types of motion: ***x-ray tube motion, image receptor motion,*** and ***patient motion.*** To achieve a sharp image, the x-ray tube, image receptor, and patient must remain motionless during the exposure.

Because of the high quality x-ray tubes and image receptors that are being manufactured today, their effects on image sharpness are negligible. Of all factors, *patient motion has the greatest effect on image blurring.* When longer exposure times are used, patient motion is the greatest factor that will affect the sharpness of an image.

Unfortunately, it is often very difficult to keep a patient perfectly stationary during long exposures for several reasons. First of all, the patient must breathe! The mechanical process of breathing can cause a significant amount of patient motion which will produce a blurry image. Other involuntary patient motion includes: heart motion, swallowing, and motion related to the physical condition of the patient. Patient restraints (i.e., straps that hold the patient in place) are only effective to a small degree. *By increasing the mA setting, a shorter exposure time can be used thereby reducing the effects caused by patient motion.*

A disadvantage of using a higher mA setting relates to the corresponding increase in the focal spot size caused by a phenomena termed ***blooming.*** Blooming is a condition that results from the larger space charge at surrounds the filament during a high mA exposure. The larger space charge causes an increase in the effective focal spot size. The effects of blooming are critical in studies where extremely sharp detail is required (i.e., angiography) and for this reason x-ray tubes with very small focal spot sizes are most often used.

The X-ray Tube

Four characteristics of x-ray tubes have a significant effect on image quality. These are: ***the focal spot size*** of the tube, the amount of ***inherent filtration*** of the tube, the type of ***anode construction,*** and the ***heel***

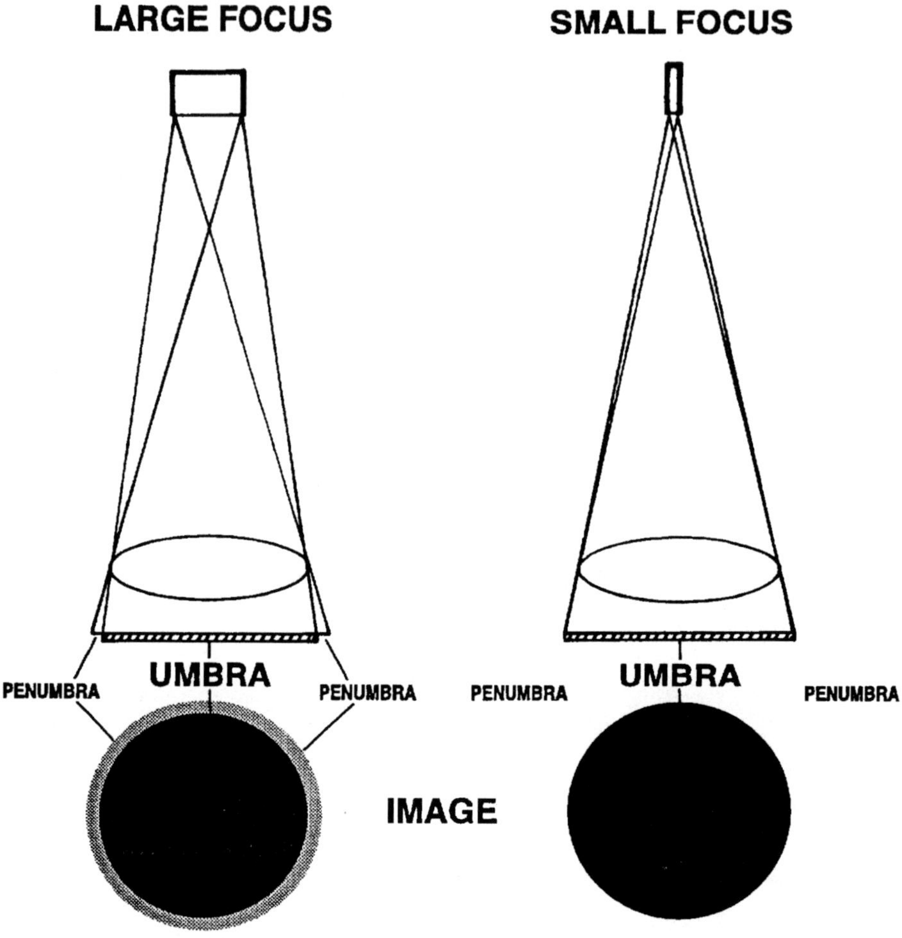

Figure 29. The effect of focal spot size. A smaller focus will produce less penumbra and therefore a sharper image.

effect produced by the anode. Each of these four factors must be considered when evaluating the images produced from an x-ray system.

The focal spot size (or focus) of the tube directly affects the amount of detail (or sharpness) that can be observed in the image. When discussing the focus, we are actually referring to the ***effective focal spot size*** of the x-ray tube. As stated in Chapter I, the effective focal spot size is dependent on a size of the filaments and the target angle. The diagram in Figure 29 demonstrates the influence of focal spot size on image sharpness. The "shadow" created from an object exposed by a point source has two components: the ***umbra,*** which is the main shadow created by the object, and the ***penumbra,*** which is the lighter blurry area surrounding the umbra.

A smaller effective focal spot size produces less penumbra and therefore greater sharpness. However, because of the reduced exposure rating capabilities it is not practical to always use small focus and, as a result, dual focus tubes are widely used. With a dual focus x-ray tube, small focus is used solely for studies requiring fine detailed images (or in most pediatric studies) and large focus is used for most other studies.

The focus of the x-ray tube is precisely measured at the tube loading facility before it is shipped to the end user. In the field, x-ray tube focus is most commonly measured by service engineers with a test resolution phantom called a ***lead test star*** at the time of installation (see Chapter VI).

With today's high standards used in manufacturing x-ray tubes, the focal spot size of the tube should remain relatively constant for the expected life of the tube. If, however, the focus of the tube is in question, it should be checked with the lead test star. *Ideally, the focal spot size of the x-ray tube should be measured when a new tube is installed or when a loss of detail is noticed on the film.*

The second characteristic of x-ray tubes that affects image quality is the inherent filtration of the tube. Inherent filtration refers to the amount of filtration that is present within the x-ray tube assembly alone. Because of their construction, all tubes have a minimum amount of ***inherent filtration.*** Specifically, this filtration is a result of the combination of the glass insert, the insulating oil, and the tube housing port, all of which are in the path of the x-ray beam.

The amount of filtration created by these materials is measured and recorded in equivalent millimeters of aluminum at the factory. This numerical value can be found in the x-ray tube data sheet that is supplied with the tube.

Since that amount of filtration affects the quality of the x-ray beam and, thus, image contrast, it is important to know the amount of inherent filtration of the particular x-ray tube. In fact, the inherent filtration must be considered in the total measurement of filtration in the x-ray beam.

With x-ray tubes, the inherent filtration will increase as the tube ages. This is due to the accumulation of tungsten that is deposited on the inside of the glass insert as a result of filament evaporation. The increase in filtration can be compensated for by removing the appropriate amount of filtration located in the collimator (see below). When the inherent filtration of an x-ray tube becomes excessive, the tube must be replaced. Usually the tube will fail in another manner before the filtration reaches this point.

The third characteristic of x-ray tubes that directly affects image quality is ***anode construction.*** The material used in the target will dictate the type of radiation that will be produced. Tungsten is most commonly used in x-ray tubes for general radiography because it produces predominately higher energy x-ray photons. Molybdenum is used in mammography because it creates much lower energy radiation which is necessary for imaging soft tissue. Today most manufacturers of x-ray tubes utilize a combination of several different materials to achieve the desired beam quality.

A final characteristic of x-ray tubes that affects image quality involves the ***heel effect*** caused by the anode. As stated in Chapter I, x-rays disperse from the target in all directions. X-rays that travel *in the direction of the anode* are partially absorbed as they pass through the thickness of the target. This absorption will cause a gradual reduction in the amount of radiation in the beam when measuring from the center of the beam towards the anode end of the x-ray tube. The gradual reduction in radiation will result in a gradual change in density that is clearly visible on a radiograph. Also, the heel effect increases with smaller target angles.

To compensate for the heel effect, the service engineer will orient the x-ray tube with the anode in the direction of the thinner anatomy of the patient, usually at the head end of the table (and patient). In this orientation, the heel effect is negligible for most common studies and therefore will not degrade image quality.

The Collimator

The collimator contributes to image quality in several important ways. The primary function of the collimator is to ***reduce the size of the radiations field*** so that only the area of interest will be exposed. This action will result not only in a significant reduction in the total dose that the patient receives but will also *significantly reduce the amount of scatter radiation created by the interaction of x-rays with the patient.* Scatter radiation causes an increase in ***fogging*** on the film and is the greatest contributor to poor image quality.

In addition, the use of collimators (also cones and diaphragms) effectively decreases the overall film density and will also increase image contrast. The collimator also removes off-target radiation, termed ***stem radiation.*** Stem radiation occurs when the electrons from the filament hit the target in areas other than the focal track. This produces off-focused radiation that will severely affect image sharpness. The upper set of blades and the "fingers" in the collimator assembly are used to remove stem radiation.

Another advantage of using a collimator is that it functions effectively to reduce the ***backscatter radiation*** originating from behind the image plane. Backscatter radiation is usually caused when the x-ray beam interacts with the radiographic table. If the blades of the collimator are closed down to the area of interest only (i.e., the specific anatomy of the patient), less radiation will be available to interact with the surrounding patient structures and the table, thus greatly reducing the amount of backscatter radiation.

Finally, an important function of the collimator regarding image quality is that it provides beam ***filtration.*** Located within all collimators is a slot that will allow for the placement of the additional filters needed to achieve the correct beam quality as specified by federal guidelines. The collimator may be equipped with selectable aluminum filters, or there will be a slot provided into which various amounts of aluminum sheets can be placed. In addition, the x-ray beam must also pass directly through the mirror, found in all collimators, which also provides a small amount of filtration.

The appropriate amount of filtration that is needed is determined by performing the HVL measurements, and then the correct thickness of aluminum is inserted in the collimator. The reason the filtration is added at this point in the beam pathway is that any defects in the aluminum will effectively be "blurred out" and will not be visible on the final image. Consequently, all filters should be placed as close to the tube as possible.

SID and OID

When positioning a patient for a radiograph, there are two spatial variables that have a great effect on image quality: the ***source to image distance,*** or ***SID,*** and the ***object to image distance,*** or ***OID.*** The service engineer must completely understand the effects caused by these geometric variables.

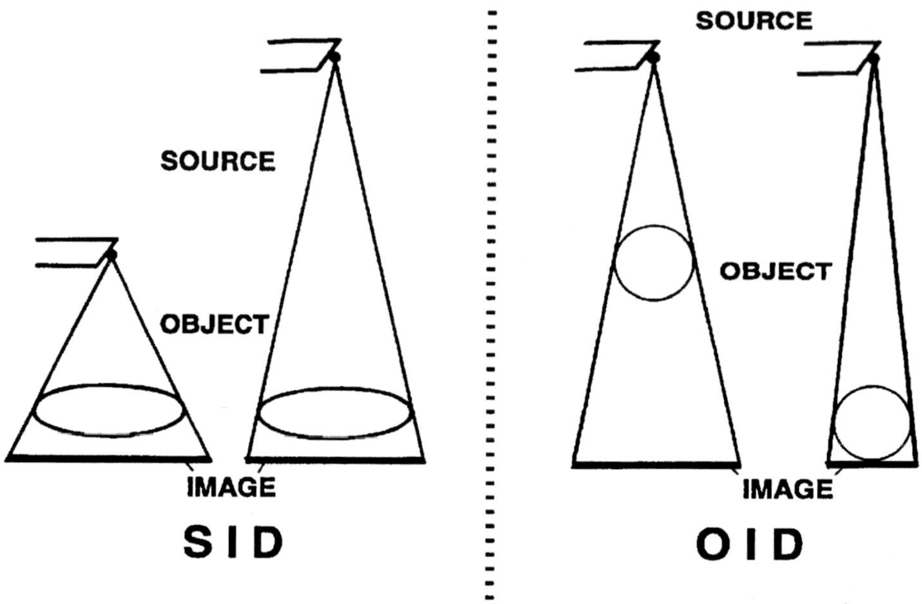

Figure 30. The effects of SID and OID. (Left): Increasing the SID will improve image sharpness and reduce image magnification. (Right): Decreasing the OID will improve image sharpness and reduce image magnification.

The SID, also referred to as FFD (Focus-Film Distance), greatly affects image sharpness. SID refers to the distance from *the focal spot of the x-ray tube to the film plane.* The diagram in Figure 30 shows the effects of varying the SID. As the SID increases, the focal spot appears to get smaller as viewed from the film plane. The effect of the smaller focal spot reduces the penumbra and improves image sharpness. Specifically, if choosing between the two most common SID's, 40 inches (100 cm) or 72 inches (180 cm), a 72-inch SID will produce a sharper image.

Changes in SID will also affect the film density. As the SID is increased, the x-rays must travel a greater distance to reach the film. Since x-ray radiation must obey the ***Inverse Square Law,*** a significant reduction in radiation occurs at greater SIDs. This reduction must be compensated for by increasing the radiographic technique. Similarly, reducing the SID causes an increase in film density, all other factors held constant. A technique that produces a good image density at a 72″ SID will produce a much darker image at 40″ SID. Using the inverse square law for this example:

$$I/i = d^2/D^2$$

where: I is the new exposure rate
i is the original exposure rate
d is the original SID
D is the new SID
$I = (72^2/40^2) \times (i)$
$I = 3.24 \times (i)$

From the equation, it is seen that by reducing the SID to 40″, the exposure rate is exactly 3.24 times greater than the original exposure rate at 72″ SID.

Finally, reducing the SID also causes ***image distortion.*** Image distortion is defined as *the misrepresentation of the true size and shape of an object.* Size distortion is referred to as ***magnification.*** When the SID is decreased, image magnification increases. ***Shape distortion,*** the second type of image distortion, occurs as a result of improper alignment of the tube, patient, or film.

The second spatial factor that affects image quality is the *OID* (object to image distance), also referred to as *OFD* (object-film distance). OID refers to the distance between the patient and the film. From the diagram (Figure 30), it is clear that an increase in OID will result in a decrease in film sharpness. Moreover, an increase in OID causes significant image magnification which is undesirable for most studies. For the above reasons, positioning the patient as close as possible to the film is the recommended practice.

For general radiography, two standard SID's are commonly used for most x-ray exams. An SID of 72″ is preferred when a sharp image with minimal magnification and distortion is desired. A chest film, for example, is usually taken at 72″ SID. If the chest film was exposed at a 40″ SID, the much enlarged image of the heart could possibly mask a lesion in the lung. The 40″ SID is, however, used for most other studies in general radiography. This distance was found to produce good imaging, especially for cranial and extremity radiography.

The Patient

In addition to the effects produced by radiographic equipment, another factor that directly affects image quality is the ***patient*** who is being filmed. We know from the preceding section that patient motion will cause a loss of detail because of image blurring and that by reducing the exposure time, the effects of patient motion will be reduced. It is the patient's internal structures, however, that will ultimately dictate how the final image will appear. Factors such as size, age, and pathology of a patient will greatly change the appearance of the final image. For this reason, the engineer must be familiar with patient morphology, standard filming techniques, and some of the common disorders that radiography helps diagnose.

Tissue will absorb radiation in different amounts depending on the atomic number of the material and the density of the tissue. Bone, for example, absorbs significantly more radiation than does soft tissue. It is exactly this tissue specific variation of attenuation that creates an exit beam that produces an image consisting of a wide range of densities on the film (i.e., good film contrast).[23]

The subject contrast varies greatly with patient type, age, physical condition, and patient pathology and must be considered when evaluating image quality. If the technique factors and geometrical setup are held constant, subject contrast will vary significantly between a large patient and a thin one. The larger patient will absorb much more radiation than the thin patient. In addition, there will also be more magnification of the internal structures of a larger patient because the organs are farther away from the film. Finally, a larger patient will produce more scatter radiation which, in turn, reduces image quality.

The age of the patient also has an effect on the image. The muscle tissue of an elderly patient will produce an image that is drastically different from that of a young healthy patient. Bone density also changes with age and must be considered when evaluating an image. An osteoporotic patient will require a much lower exposure technique than a patient with normal bone tissue.

The physical condition of the patient will also affect image contrast and density and, therefore, should be considered when evaluating images. For example, a patient with lung disease will produce an image that will differ greatly in appearance, depending on the type of disease. An image produced from a person with asbestosis will differ greatly from that of a person with emphysema. Finally, the image quality will be greatly affected by a patient wearing a plaster cast or some type of prosthetic device.

---

23. If the radiologist must view a structure that is surrounded by tissue of the same density, a contrast medium such as barium sulfate or iodine is used to artificially create a high contrasting area on the film.

The Radiographic Table

The radiographic table will also have a significant effect on the final image. Here, the ***tabletop,*** which actually supports the patient, is positioned directly between a patient and the image receptor–directly in the image pathway. Consequently, the tabletop must be comprised of materials that will not attenuate the x-ray beam and, thereby, cause a reduction in image quality. Technically, the tabletop should have high transmission and low absorption characteristics.

In addition, the tabletop must be homogeneous and free from defects that could produce image artifacts. In the past, tabletops were made mostly out of wood. Unfortunately, any defects in the wood, or the wood grain itself, will show up on the image. Most tabletops manufactured today are made from carbon fiber materials and, with rare exceptions, meet all of the requirements stated above.

The Image Receptor

The image receptor, simply defined, is a detector that converts the invisible image to a visible one. When discussing the image receptor in a radiographic room, we refer specifically to the ***film*** and ***film cassettes.*** We will also include a discussion of the ***x-ray grid*** along with the image receptor since it is usually located within the image receptor cabinet. As a rule, an x-ray grid is used for most x-ray exams with few exceptions–as with tabletop radiographs, for example. Each of the components of the image receptor will have a significant effect on image quality.

The image receptor in a general radiographic room can be found in two places. The receptor located on a wall is used for erect studies. With the wall receptor, the patient is positioned so that he or she is standing directly in front of the image receptor, directly in the path of the x-ray beam. The other location of the image receptor is in the radiographic table, just beneath the tabletop.

The Grid

As the x-ray beam exits from the patient, it must first travel through the x-ray grid before it reaches the film cassette. *The sole function of the grid is to reduce the amount of scatter radiation caused by the interaction of the x-ray beam with the patient.* By reducing the amount of scatter radiation, image contrast is greatly improved.

For general radiography, both ***stationary*** and ***reciprocating*** grids are used. A stationary grid remains motionless during the exposure and is usually located within the ***grid cabinet.*** Stationary grids are commonly used for general radiography but are somewhat limited in their ability to reduce scatter radiation. Reciprocating grids, located in a ***bucky,*** will oscillate back and forth during the exposure. A reciprocating grid is more efficient at reducing scatter radiation than a stationary grid, but requires additional circuitry to create the grid movement. The added circuitry increases the overall cost of the receptor and also has an increased potential for failure.

Regardless of the type of grid, they can be very costly mostly because of their special construction. Both stationary and reciprocating grids are made of thin ***lead strips*** separated by an ***interspace material.*** This interspace material must have a low absorption quality in order to allow the x-ray photons to pass through unattenuated. Common interspace materials are fiber, aluminum, and air.

The lead strips can be oriented parallel to each other, as in a ***linear grid,*** or the strips can form a cross-hatch pattern as in a ***cross grid*** (Figure 31). Cross-hatched grids are more effective in reducing scatter than linear grids, but are extremely sensitive to x-ray tube positioning.

The most common type of grid used today is the linear grid because it allows for tube angulation for special radiographic views without causing *grid cut-off* (see below). Linear grids can be used effectively if the x-ray tube is angled along a path which is parallel to the lead strips. For this reason, the grid axis, indicated on the front (tube side) of the grid, must always be oriented in the direction of the tube angulation (head to foot direction of patient).

With both types of grids, the lead strips are spaced to allow most of the primary beam to pass directly through to the film, while impeding any scatter radiation traveling at angles to the film plane (Figure 32). Even so, some of the primary beam will also be impeded by the lead strips, reducing the overall amount of radiation hitting the film. Consequently, the x-ray technique must be increased when using grids.

In fact, all grids are designed to operate at a specific range of kilovoltage. If used out of this range, density variations will occur on the film. The specified kilovoltage is directly dependent on the amount of lead in the grid. Consequently, x-ray grids with high grid ratio's and grid frequency's (see below) require higher kilovoltages.

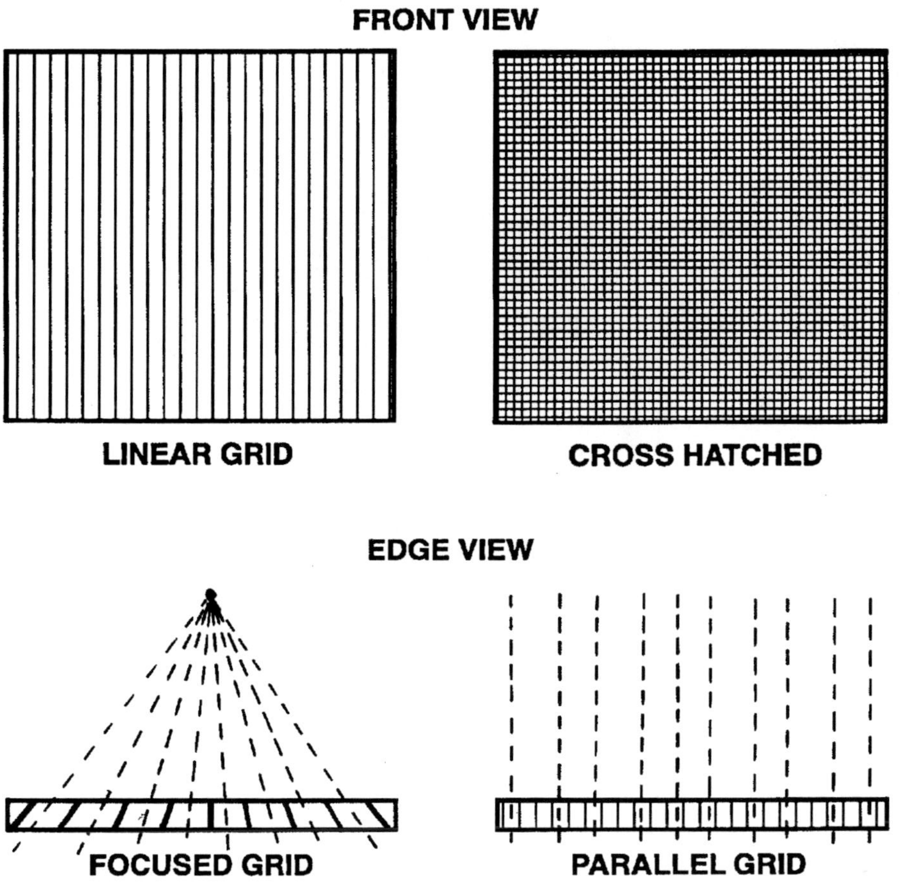

Figure 31. Types of x-ray grids.

The two major factors that determine how effective a grid is at reducing scatter radiation are the ***grid ratio*** and ***grid frequency.*** The grid ratio is the *ratio of the height of the lead strips to the distance between each lead strip.* A grid having lead strips that are 0.08 inches high, separated by 0.01 inches, has a grid ratio of 8 to 1. A higher grid ratio is more efficient at reducing scatter radiation.

With higher grid ratios, there is more lead for the radiation to pass through, and consequently, a higher exposure technique is required to achieve an adequate film density. Also, grids that have high grid ratios are more sensitive to positioning than grids with lower ratios. If the tube is not perfectly aligned with the receptor, ***grid cutoff*** will result. It is for this reason that stationary grids with low grid ratios (e.g., 5 to 1) are used with portable x-ray machines.

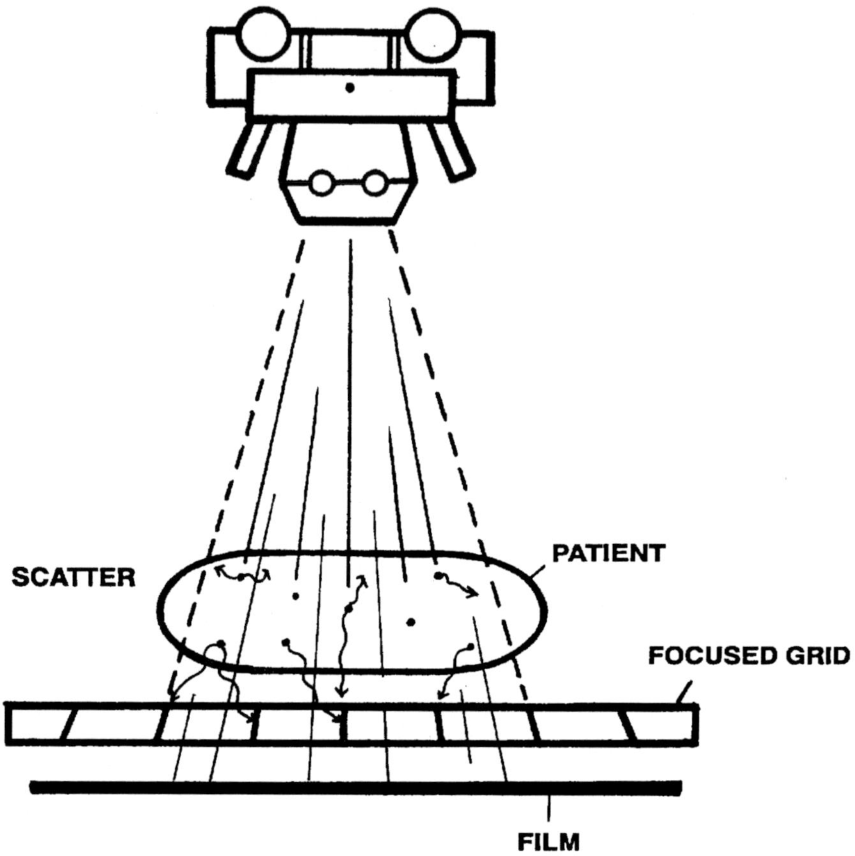

Figure 32. How the x-ray grid removes scatter radiation. Only radiation traveling in straight lines will reach the film.

The grid frequency refers to the number of lead strips that are in the grid and is measured in grid lines per inch (or lines per cm). As the grid frequency is increased, the width of the lead strips will decrease to accommodate for the additional strips (while maintaining the same grid ratio). The thinner strips of lead found in a grid with a high grid frequency will not be visible on the radiograph. A grid with thicker lead strips (or low grid frequency) will greatly reduce scatter radiation but the image of the lead strips will appear on the film as ***grid lines.***

When selecting a grid, a compromise must always be made between grid frequency and the amount of scatter reduction desired. Common grid frequencies used in radiographic rooms are in the range of 80 to 100 lines per inch. At these grid frequencies, the x-ray grid can effectively reduce scatter radiation without causing visible grid lines on the image.

X-ray grids can be either of ***focused*** or ***parallel*** construction, depending on how the lead stripes are oriented in respect to the axis of the x-ray beam. A focused grid has lead strips that are angled as they move away from the center of the x-ray beam. If lines were drawn through the angled strips, they would converge at point above the film plane (Figure 31). The point where these lines converge is the focus point of the grid. Focused grids can be focused for a specific SID such as 40″, for example, or for an SID range such as 40″ to 72″ SID. The x-ray tube must be positioned at the point of focus or ***grid cutoff*** will result. This positioning is very critical with higher grid ratios.

With parallel grids, the lead strips are not angled but all run parallel to the central ray of the x-ray tube. Parallel grids must be used with a large SID and small film sizes or grid cutoff will result. Parallel grids are often used for spot filming in fluoro suites.

Grid cutoff is caused when a significant portion of the primary x-ray beam is stopped by the lead strips. By impeding part of the primary beam, less radiation reaches the film and, consequently, a decrease in film density can be observed on all (or part) of the radiograph. There are four types of grid cutoff, and each type produces an effect on the final image. The service engineer must understand the causes for each type of grid cutoff and the effects they produce on the film.

1. ***Off-center cutoff*** is a result of the x-ray beam not being centered to the grid. This type of cutoff produces a general decrease in density over the entire film.
2. ***Off-focus cutoff*** is caused by having the x-ray tube positioned out of the focal range of a focused grid. This type of cutoff produces lighter densities on the side borders of the film while the center remains unaffected.
3. ***Off-level cutoff*** when the grid is tilted in respect to the film plane and is seen as a general decrease in overall density of the film.
4. ***Reverse cutoff*** is caused by a grid that was positioned upside-down in the path. This type of cutoff results in a decrease in density around all borders of the film.

The Reciprocating Bucky

The grid used in a reciprocating bucky can have a very low grid frequency (i.e., thicker lead strips) without causing grid lines. The reason for this is because the bucky contains the additional circuitry that will "oscillate" the grid. The grid "oscillates" (or reciprocates) in the cross-bucky (lateral) direction, at right angles to the lead strips. By doing so, this cross motion effectively blurs the thicker lead stripes so that they will not be seen on film. As stated earlier in this chapter, grids with a low grid frequency are more effective at reducing scatter.

Reciprocating buckys require a motor and drive assembly to cause the grid to oscillate. Also, the grid must move at the ***correct speed*** and must be ***synchronized*** with the generator pulses or grid lines will result. The speed and synchronization of the bucky motor are controlled by the x-ray generator. The effects of motor drive problems are especially noticeable on very short exposures.

The Film Cassette

The introduction of the film cassette marked a major advancement in general radiography and is, today, considered an essential part of the image formation process. However, since the film cassette lies directly in the path of the x-ray beam, it will significantly affect image quality.

The film cassette (Figure 33) is basically a ***light-tight container*** that holds the unexposed x-ray film during the exposure. The cassette is inserted into a cassette tray, cassette holder, or it can be placed directly behind the patient when filming. Film cassettes are available for every standard film size used in general radiography and are compatible with all makes of x-ray equipment.

The most significant feature of the film cassette is that it contains ***intensifying screens.*** The sole function of the intensifying screen is to convert the x-ray energy to light energy which will then expose the film. In fact, 98 percent of the energy that exposes the film is light energy. X-rays contribute only about 2 percent of the total exposure. By converting x-ray radiation to light energy, the light sensitive film is exposed much more quickly than if exposed with direct x-ray radiation. The reduction in exposure time not only reduces the radiation to the patient but also extends the life of the x-ray tube.

Unfortunately, the best screens available can never match the resolution of a radiograph that was produced without intensifying screens. However, the benefits of using intensifying screens far outweigh this one drawback and, today, all diagnostic radiographs are taken with the intensifying screens.

The intensifying screen consists of a layer of phosphor crystals supported by a cardboard or plastic base. Since ***double-emulsion film*** (i.e.,

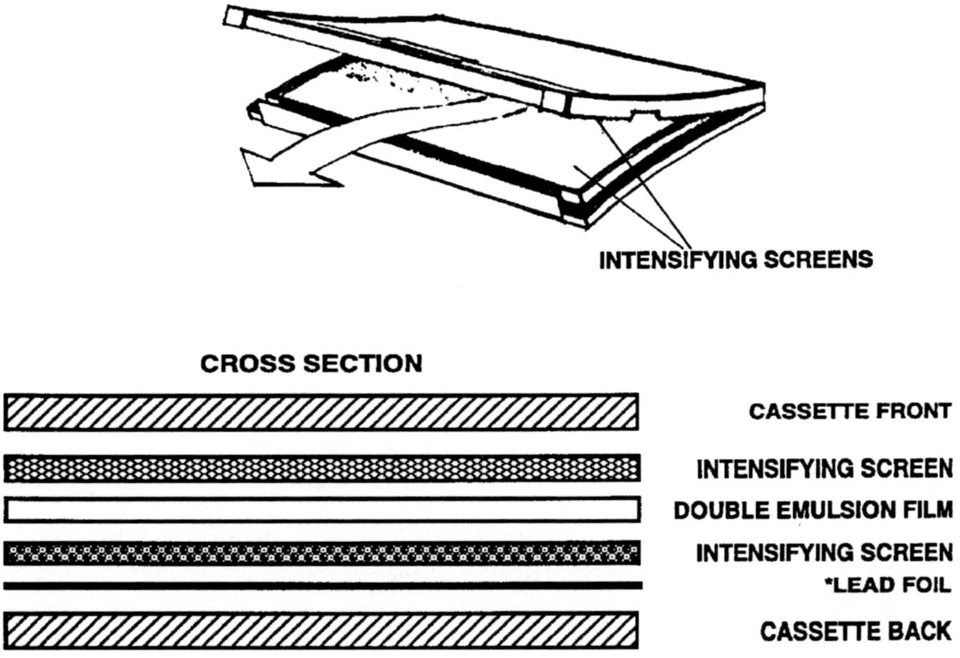

Figure 33. The film cassette. The lead foil is used to reduce backscatter radiation.

an emulsion layer on both sides of the film) is used in general radiography, two intensifying screens are used inside the film cassette. The film is "sandwiched" between the upper and lower screens and is held firmly against each intensifying screen by ***springs*** that are also located within the cassette. Without the spring tension, the insufficient ***film/screen contact*** would cause a blurred image. Good film/screen contact is needed to produce sharp images.

The phosphor crystals emit light energy when exposed to x-ray photons, and this light energy then exposes the film. The most common phosphor crystal used in the intensifying screens is ***calcium tungstate*** which emits light in the blue and ultraviolet region of the electromagnetic spectrum. Calcium tungstate is used because there is very little lag (i.e., has a quick response time) and it is very durable as a screen material. Today, rare earth phosphors are also used because of their greater efficiency in converting radiation to light energy. Rare earth phosphors such as gadolinium and lanthanum emit light in the green region of the visible spectrum and must be appropriately matched with green sensitive film.

When discussing intensifying screens, the most important characteristic is ***screen sensitivity.*** The size of the phosphor crystals and the thickness of the phosphor layer determine the sensitivity of the intensifying screens. A screen that contains larger crystals or that has a thick layer of phosphor will give off significantly more light energy for a given amount of radiation than will screens with smaller crystals (or thin phosphor layer).

The increased light output of the larger crystals will result in much shorter exposure times. These screens are termed ***fast*** screens or ***high sensitivity*** screens. However, larger crystals will no resolve the fine detail of certain anatomical structures because of the increased amount of light output. *A fast screen has large crystals and produces less detail.*

Where extreme detail is needed, a "slower," less sensitive screen is needed–one with smaller crystals and a thin phosphor layer. These screens, called ***detail*** screens, require a higher exposure rate to produce a density equal to that of the fast screens. *Detail screens have small crystals which provide better detail.* Detail screens are mostly used for extremity work and angiography.

X-ray Film

In general radiography, the image media where the final image is formed and permanently stored is x-ray film. There are two types of x-ray film commonly used today. ***Double emulsion film*** is used for most radiographic studies and consists of a base material, made of plastic, with an emulsion layer on both sides of the base. Single ***emulsion film*** has an emulsion layer on one side only and is used for special applications such as mammography. The emulsion contains crystals of silver bromide that are suspended in a gelatin mixture. The crystals, when exposed to light, will undergo a chemical change that causes neutral silver to be deposited on the film in place of the exposed silver bromide. The process of depositing silver grains forms what is termed the ***latent image.***

The size of the silver bromide crystals determines the ***film speed*** or sensitivity. A fast film consists of much larger crystals than a slow film. Since fewer (large) crystals are required to produce the image, fast film has more graininess and appears "noisy." Detail film has many more smaller crystals which produce much less grain on the film.

As mentioned in the preceding section, the correct film type must be used with a given set of intensifying screens to achieve optimum image quality. Matching a film to a particular screen is referred to as the ***film/screen combination.*** Since x-ray film can be sensitive to different wavelengths of light (i.e., blue or green light), a mismatch in film/screen combination could result in an unexposed film or other density problems. This is often a source of problems in hospitals where several different types of film are used. The service engineer must observe and note the type of film and screen being used when diagnosing imaging problems.

The film/screen combination will also affect the amount of ***radiographic mottle,*** or "noise," that will be visible in the final image. Radiographic mottle is the unwanted density variations on the film that can obscure the visibility of the actual image. In addition, radiographic noise is also generated from four other sources: ***film graininess, screen mottle, film artifact,*** and ***quantum mottle.***

Film graininess is a result of the clumping of silver bromide crystals. Screen mottle is caused by variations in the output of the phosphor crystals that comprise the intensifying screens. With the high quality film and screens of today, these two sources do not contribute significantly to radiographic noise. Furthermore, film artifact, caused by mishandling of the film or problems in the film processor, is easily identified and not a major contributor to noise problems.

Of all the sources of radiographic noise, quantum mottle is the greatest contributor. Quantum mottle is defined as the density variations on the film caused by x-ray "packets" of energy. X-rays are known to travel through air in two ways: in waves of energy, as described in the wave theory, and in "packets" of energy, as described in the particle theory. The packets of energy, referred to as ***quanta,*** are spatially distributed in a random pattern when they travel through air. As the packets strike the intensifying screens, the randomness of the radiation causes minor fluctuations in the intensity of the light emitted from the screens. The fluctuation of light, in turn, produces a corresponding fluctuation in density on the film and, hence, the quantum mottle seen on the film.

The quantum mottle of a radiograph will be more visible with fast film/screen combinations because less radiation (i.e., fewer photons) is required to create the image. Fewer photons will produce fewer black specs on the film that will be visible on the image. Using a fast film/screen combination is, in effect, creating a high gain imaging system. High gain is another word for *high amplification.* The few number of photons (energy packets) used to expose the radiograph will be amplified, and become clearly visible on the film.

A good example of this phenomenon can be easily demonstrated with a dot matrix printer. A 9-pin print head will produce characters that will appear generally faint and unsharp on paper. Upon closer examination, the individual dots can easily be discerned. A 24-pin printer, however, produces a much darker, sharp character, with individual dots that are not so readily apparent.

The exact specifications for each type of x-ray film can be graphically shown with the ***characteristic curve,*** or ***H & D curve,*** named after Hurter and Driffield who first used the curves. The characteristic curve is created by plotting a graph of the film density vs. the logarithm of relative exposure for the film and processing being used. From the curve, the ***film speed, gross fog,*** and ***film contrast*** can be determined (see Figure 34).

The curve of the film consists of a toe, straight portion, and shoulder. The best film quality is obtained with exposures falling in the straight portion of the curve. The slope of the straight portion, called the ***average gradient,*** defines the overall film contrast. Exposures that fall outside the range of the straight portion will have reduced contrast. This is the reason why it is difficult to see density differences on a film with a density above 2.5 D or below 0.5 D.

From a service standpoint, the film type must be noted when evaluating image quality. In addition, the engineer should check that the film is correctly matched to the intensifying screens (film cassettes) used so that the correct film density and optimal image quality can be achieved. Since the film speed is clearly marked on the packages of film and the intensifying screen speed is usually indicated on the cassette, the service engineer need simply verify that they are correctly matched.

In addition, the film should be stored at the proper temperature (60°F–80°F) and at the correct humidity (30%–50% R.H.) to ensure that it remains "fresh" (i.e., in good condition). Also, all film has an expiration date, clearly labeled on the film box, that indicates the useful life of the film. Again the engineer should check these areas first when diagnosing imaging problems.

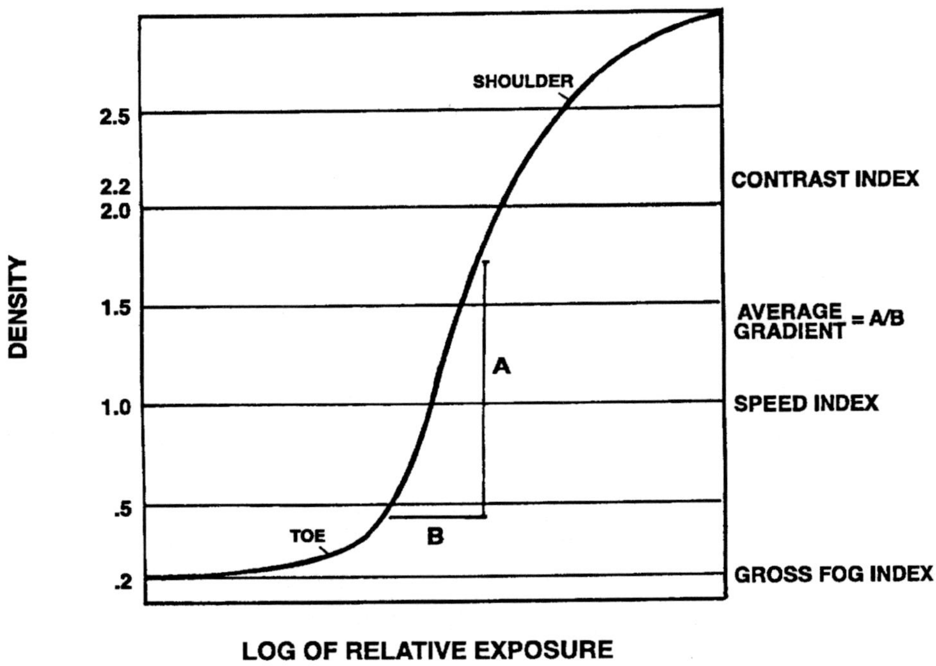

Figure 34. The H & D curve.

Film Processing

Another major factor that can affect image quality is film processing. Perfectly calibrated x-ray equipment will be of little use if the processor is not working properly. *Without a properly functioning film processor, it will be impossible to get a good radiograph.* The film processor, in this sense, is the "weak link" of the imaging chain and, consequently, is the cause of many imaging problems.

Although the x-ray service engineer need not be an expert on servicing film processors, he or she must have a solid understanding of the film development process and the effects that improper film processing has on image quality. This is important since many service calls occur after normal working hours when no one is available to confirm the proper operation of the processor. Thus, the service engineer should feel comfortable with performing minor checks on the film processor in order to rule out "processing" as the cause of imaging problems. Since the processor is fairly straightforward in design, it should be one of the first areas that is checked when troubleshooting imaging problems.

After the film has been exposed, it is taken into a ***darkroom*** where it will be removed from the film cassette and inserted into an automatic film processor.[24] The darkroom plays an important role in image quality in that any *light leaks* in the darkroom will cause fogging on the film which will severely degrade image quality. In addition, the *safe-light,* found in all darkrooms, must contain the correct optical filter for the type of film being used. The light bulb inside the safe-light must also be the correct wattage, and the light must be at the proper distance from the film loading area or film fogging could result.

Many radiology departments today use an automatic daylight film unloading/reloading unit in conjunction with the film processor. This unit eliminates the need for a darkroom thereby eliminating all the darkroom-related problems. With these automatic units, the film cassette is inserted into the cassette loading section where it is automatically pulled into the machine, and then opened in a light tight area inside of the unit. The film is removed from the cassette and then automatically loaded into the processor section. Finally, the cassette is reloaded with a fresh film and then ejected back out to the loading slot.

Once the film enters the automatic film processor, it travels through a series of rollers to the developer tank, fixer tank, the water tank, and dryer, respectively (see Figure 35). The developer forms the image on the film by reducing the neutral silver deposits on the film to black metallic silver. When the film leaves the developer tank, the complete image will be visible on the film. The remaining stages of film processing will just make this image permanent. For this reason, the developing stage is the most important stage of film processing. Any problems encountered at this point will remain unchangeable.

The developer must be active (i.e., "fresh") and maintained at the specified ***temperature*** or image quality will be severely affected. In fact, the temperature of the developer is probably the most important factor in film processing. A variation of a few degrees in either direction will cause significant changes in film contrast. In addition, the film must remain in the developer for a specific length of time, called the ***immersion time,*** or improper development will occur. Finally, the developer must be replaced (i.e., ***replenished***) at a constant rate ***(replenishment rate)*** to maintain the correct amount of reducing agent.

---

24. Today, most facilities use automatic film processors because of the significant amount of time saved (i.e., 90 seconds for automatic processing vs. 20 minutes processing by hand).

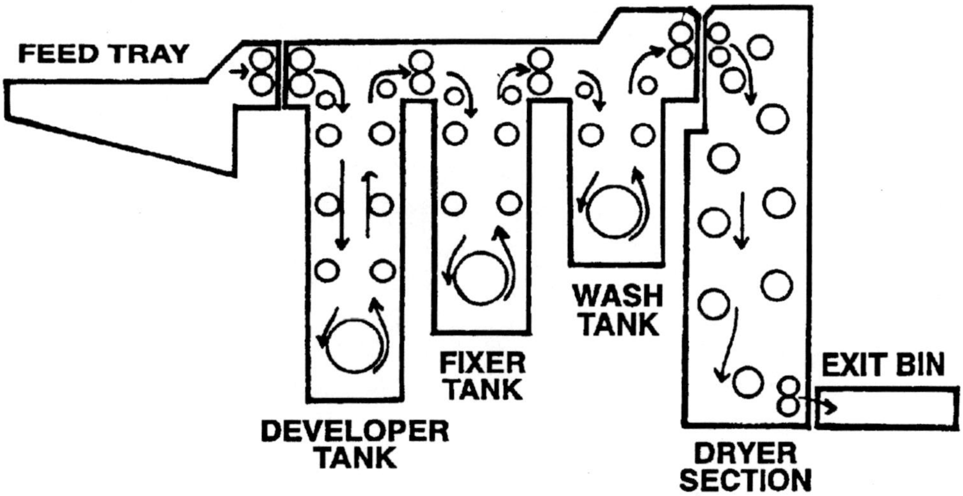

Figure 35. The automatic film processor.

When the film travels to the fixer tank several things occur. The fixer, usually made of ammonium thiosulfate, chemically stops the action of the developer and permanently ***fixes*** the image on the film. It does so by removing all the unaltered silver bromide crystals that were not exposed to light and, in effect, ***clears*** the film. If these unaltered crystals were left on the film, they would react with ambient light and cause a severe loss of image quality over time. The fixer must always remain fresh or it will not completely inhibit the developer. The fixer also contains hardeners that protect the film from scratches that may occur from routine handling.

When the film has finished traveling through the fixer tank, it is then rinsed thoroughly in the wash tank. All the unused chemicals are now removed from the film so that no other reactions will occur that could affect image stability. In addition, the film is washed to eliminate ***film artifacts,*** such as spotting. Water filters are usually inserted in the water supply line to eliminate film artifacts caused by rust or salts in the water.

The dryer functions to completely dry the film after it has been rinsed. If the films are not completely dry when exiting the film processor, they will stick together in the holding tray at the output side of the processor and, as a result, will be useless for diagnosis.

The time that the film remains in each tank (i.e., immersion time) is directly related to the speed of the automatic processor. The immersion time is directly related to the speed of the motor that drives the processor rollers. This motor is adjusted according to the strength and the temperature of the developer and fixer. In the automatic processors of today, these chemicals are mixed to extremely strong solutions. In addition, the temperature is usually held at a very high setting, 94°F, for example. This allows for extremely fast processing of about 90 seconds, which is a significant improvement over hand processing.

Film processing should always be monitored with a **quality control program.** This program ensures optimum image quality by monitoring the consistency of the film processing. Performing film processor quality control takes only a few minutes and should be done on a weekly basis. The procedure requires that a test film be exposed with a "step wedge" of increasing densities. The device used to create the step wedge, the ***sensitometer,*** produces a series of exposures on the film having known ratios to each other. The sensitometer has a calibrated lamp, and a timer circuit that precisely exposes the film. The film is then "read" with a ***densitometer*** which displays the logarithmic value called film density.

A step midrange on the strip (i.e., at 1.2 D) is used as the ***speed index.*** This step is charted on a graph and is used to track the speed of the processor. By doing so, any major variations can be observed on the graph and then the appropriate corrective action can be taken. In addition, a densitometer reading is also taken a few steps above the speed step (near 2.2 D). The ratio of this reading ***(contrast index)*** to the speed index is charted on the graph and is used to indicate the film contrast. By viewing the contrast reading on the graph, processor conditions are constantly monitored.

A third reading, called the ***gross fog*** (or ***base plus fog***) is also very important for quality control. Specifically, a clear film (i.e., an unexposed film that has been run through the film processor) will produce a reading on the densitometer that is at some value above zero. This reading is due to the material used in the film itself. Usually, the first step on the sensitometric strip is unexposed and, therefore, the gross fog measurement is made here. If the gross fog value is recorded and monitored on a regular basis, any increase in film fogging will be immediately noticeable and can be corrected before patients are examined. Usually a light leak within the darkroom or within the cassette itself is the cause of increased film fogging.

Viewing the Image

When the completely processed x-ray film emerges from the film processor, it is ready for viewing. Here is the only time in our discussion of image quality that we must introduce a ***subjective element.*** The fact is that it is the image interpreter, usually the radiologist, who will ultimately decide if the image is good or not. An image that would be considered good in the eyes of a service engineer may not be acceptable at all in the eyes of the radiologist. Since each individual's perception is unique to that individual, the subjective factor can be a source of aggravation when dealing with image quality problems.

For this reason, any variables that could alter the perception of the image should be eliminated. Specifically, the viewing conditions must be held at an acceptable level so as to ensure that all aspects of the image can clearly be seen without distraction. Moreover, these viewing conditions should be standardized so that a film viewed at one hospital will look the same when viewed at a different facility.

The radiologist will view or "read the film" on a ***view box*** (or ***illuminator***). An illuminator is basically a box with an opaque screen on the front side. Fluorescent lights located within the view box provide the back lighting for the screen. When the film is placed on the opaque screen it will be illuminated by the fluorescent lights.

The main purpose of the view box is to provide even (ambient) lighting, at a specific intensity and color, across the entire viewing area. A view box that provides uneven lighting or low illumination will directly affect the radiologist perception of the image. This is especially true for mammography where proper illumination becomes extremely crucial. The guidelines for view box luminance, specified in average luminance, is 40 foot-lamberts or 1400 cd/m squared.

The room lighting must also be controllable in the viewing area so that the overhead lights can be dimmed to an acceptable level or shut off completely. In addition, any sources of stray lighting should be eliminated since it could affect the appearance of the image. Technically, the room should be dark enough so that the radiologist's eyes will become ***dark adapted.*** Under these conditions the human eye can better discern minor differences in light intensity on the film.

Even with the viewing conditions held to a standard level, the subjective element (i.e., human perception) still becomes an important factor. The service engineer must learn what criteria the radiologist uses in determining what makes a good image. In fact, much time should be invested in this area because the final judgment of the radiographic equipment (and equipment service) lies in how the radiologist perceives and likes the images produced in the x-ray room. After all, the point of the installation and calibration of the radiographic equipment is to produce a good image. Ultimately, it is the radiologist's opinion of the quality of the images produced that determines the success or failure of a radiographic room.

Chapter IX

TROUBLESHOOTING AND REPAIRING RADIOGRAPHIC EQUIPMENT OVERVIEW

When "quality" radiographic equipment is properly maintained, it will provide relatively trouble-free operation for the stated useful life of the equipment.[25] Eventually, however, certain components of a radiographic system will experience failures regardless of quantity or quality of the maintenance provided. After a period of use, the mechanically moving parts of the x-ray equipment will begin to wear as a result of friction to the point where they can no longer function properly. Also, electronic components will eventually fail after being subjected to the countless heating and cooling cycles that occur with daily use and from being exposed to electrical power fluctuations.

X-ray equipment manufacturers take these factors into consideration when they establish the number of years the equipment is expected to function at an acceptable level of performance (i.e., useful life). The predictions for useful life are based on the amount of time that the equipment will be used, which, in turn, is directly related to the number of patients examined. When predicting equipment longevity, equipment manufacturers assume that the recommended maintenance will be performed regularly on the equipment.

There is, however, a major factor that can greatly alter any predictions of equipment longevity–*the human factor.* Radiographic equipment is designed to be used in a specific way. When properly operated,

---

25. The expected useful life of x-ray equipment varies among manufacturers. When properly maintained, most equipment should provide relatively trouble-free operation for 10 to 15 years. This is, of course, a broad generalization. Patient volume must also be considered and can greatly alter any predictions of equipment life.

the predicted useful life of the equipment is easily achieved. On the other hand, if the equipment is improperly handled, the life expectancy will rapidly decrease in an unpredictable way. Although equipment manufacturers account for some rough handling in their equipment designs, negligent use over a period of time will result in a high rate of mechanical failures. The plain fact is that humans do not always work well with machines and ***operator mishandling*** accounts for a significant number of service calls.

In fact, most downtime for radiographic equipment is attributed to mechanical failures. The high occurrence of mechanical failures in radiographic equipment is directly related to the daily handling of the equipment by the x-ray technicians (or radiologist). Although some mechanical parts will eventually wear out with normal handling of equipment, careless handling will significantly increase the failure rate of the equipment and, thus, shorten equipment life. Because of the sensitive nature of this topic, any discussion of operator mishandling must be dealt with in a delicate fashion to prevent personal accusations from arising (see Chapter XIII). Equipment mishandling is, however, a fact of life and the service engineer should be aware of the fact that he or she will be performing mechanical repairs on a majority of service calls.

This point can easily be demonstrated when one observes the condition of equipment in a large teaching hospital and compares it with the equipment installed in a smaller private office. The difference in appearance and operating condition of the equipment is striking. The radiographic equipment in the smaller office will invariably be in much better physical condition. This is true even if the equipment in the smaller office is much older, as is often the case.

The explanation for this wide discrepancy can be attributed to the fact that large hospitals have many different technicians, on several shifts, operating the equipment. In this situation it is often difficult to monitor equipment use or to pinpoint those who are abusers. Moreover, it is sometimes difficult to thoroughly train each individual technician on the proper use of equipment at larger medical facilities. Lack of adequate training for technicians (i.e., user inservicing) will result in the increased failure rate of equipment due to misuse.

Another reason for the wide discrepancy in equipment condition may be attributed to the fact that with larger hospitals, the person or persons who own the equipment are not usually present to observe its daily use. Consequently, technicians may feel less responsible when using the equipment.

In a small office, on the other hand, only one or two technicians may actually operate the x-ray equipment. In addition, the equipment was most likely purchased by the radiologist who owns (and operates) the office. This means that there is someone present who is concerned with how the equipment is being used, or misused. In this setting, the staff is more conscious of proper equipment handling because if it breaks, there is nowhere to pass the blame.

It could be argued that the larger hospital examines many more patients than the smaller office and this sufficiently explains the difference in condition of the equipment. The author would disagree with this argument, however. Although patient volume does indeed have a great effect of the longevity of equipment, it does not alone account for the drastic difference in the physical conditions of the equipment.

Consequently, most failures with radiographic equipment are ***mechanical*** in nature. As high as 60 percent of routine service calls involve repairs to items such as control knobs, equipment handles, cassette trays, motor drive assemblies, and roller bearing assemblies. In addition, many failures occur in electromechanical devices such as switches, relays, and electromagnetic locks. Again, any device that has moving parts generally will have a higher potential for failure. Most of these devices will fail as a result of worn or damaged parts, or from missing hardware caused by routine handling (and mishandling).

In contrast to the high rate of mechanical failure, the frequency of ***electronic*** component failure is relatively low. The high quality electronic components used on circuit boards in x-ray equipment today result in extremely reliable operation. This increased reliability is also attributed to the improved manufacturing techniques used on circuit boards. It is not uncommon to find perfectly functioning, 20 year old, x-ray generators that still contain their original (factory installed) circuit boards.

There are, however, other sources of electronic failures in radiographic equipment. One source of failure occurs with the ***equipment cables.*** This is true for all types of cables, including power cables, signal cables, and high voltage cables. Cable problems often result from the repeated overflexion of the cable that occurs during normal movement of the x-ray equipment. The constant stress applied to the cable will eventually causes the many strands of wires in the cable to fray and, finally, completely break apart.[26]

The cables that are most susceptible to stress are the interconnecting cables that run to the tube support, since it is routinely repositioned for each patient. These cables include the high voltage cables, rotor cable, collimator cables, and the cables that supply power to the electromagnetic locks. If these cables are stressed during normal tube movement, it is an indication that they were not draped correctly at the time of installation (see Chapter VI). The service engineer should, therefore, redrape these cables once the repair is completed. This will ensure that the failure will not reoccur.

Other electronic failures can occur in the ***primary*** and ***secondary*** power circuits of the x-ray generator. These failures are the most difficult for x-ray service engineers to diagnose for several reasons. First of all, these types of failures do not occur on a regular basis with radiographic equipment. A primary contactor, for example, might possibly fail once in the entire life of a given x-ray generator. Consequently, a service engineer may only see this type of failure a few times in his or her career. As a result, most service engineers have little experience with this type of failure.

Second, because of the danger of electrical shock, the engineer must use great caution when troubleshooting the primary and secondary x-ray circuits. Oftentimes the failure can be detected only by the presence of high voltage arcing. Because of the potential hazards involved, special troubleshooting techniques are required when servicing these circuits to ensure that the service engineer will not be exposed to high electrical potentials. He or she must pay strict attention to the in-circuit placement of test equipment in these power circuits or damage to the equipment (or the engineer) could result. In addition, the differences between ground references and floating references must be identified in these circuits before the engineer can connect the test equipment to the individual components.

Another factor that makes troubleshooting the high power circuits difficult is that many of the individual components of these circuits cannot be tested by using standard troubleshooting methods. For example,

---

26. As a rule, every cable used in an x-ray room is made of multistranded copper. This is much preferred over the solid core cable used by electricians mainly because it is much easier to work with. Signal cables will have additional shielding provided under the rubber cover.

simple ohmmeter readings will not indicate whether a semiconductor in these circuits is defective. This is also true for testing the high voltage cables. Many of these components will only fail under a loaded condition (i.e., when high current is flowing through the component). Consequently, the troubleshooting resources of the service engineer are severely limited in this area.

To complicate matters further, the engineer should only ***attempt as few exposures as possible*** when trying to observe a high voltage failure. The reason for this is that if high voltage arcing occurs during the failure, other components in the system could be damaged. Again, the service engineer must utilize his or her ability to make quick observations so as to avoid causing further equipment damage.

Besides mechanical and electronic failures, another type of service call an engineer will normally encounter relates to ***image quality problems.*** Troubleshooting imaging problems is considered by many to be the most difficult, yet most interesting aspect of the x-ray service engineer's career. Imaging problems are caused by a failure somewhere in the imaging chain that results in a loss of image quality. Specifically, each device in a radiographic room has some effect on the quality of the final image. A failure in the imaging chain can be difficult to isolate without a thorough understanding of the ***image formation*** process (see Chapter VIII).

As stated in the previous chapter, the condition of the x-ray film, film cassettes, and film processor can also greatly affect image quality. In fact, many imaging problems are directly related to a problem with the film processor or film cassettes and not at all caused by the radiographic equipment. A key when troubleshooting image quality problems is to first rule out film and film processing as the cause.

The service engineer must apply knowledge of every aspect of radiography when diagnosing image quality problems. Since this area of x-ray servicing requires the highest expertise, it is usually performed by experienced engineers who have been working in this field for many years.

Another service-related problem is one that results from an error made by the technician who is operating the equipment. ***Operator error*** occurs when the x-ray technician is unfamiliar with the correct use of the x-ray equipment. The operator may not have been properly inserviced at the time the unit was installed, or it could be a matter of a new employee who never had the time to read the lengthy operator manuals.

Regardless of the cause, operator error accounts for a great number of service calls. This type of service call can reflect badly on the service engineer (and the equipment manufacturer) and should therefore be corrected as soon as possible. With service calls caused by operator error, the staff at the hospital will usually forget the actual cause of the service call and, consequently, note only that the equipment is always "down" and the service engineer is regularly at the site.

Operator error is, however, easily identified and the engineer will usually have the x-ray room up and running within a few minutes. For example, the problem may be caused by a switch that has been incorrectly depressed at the console placing the equipment in the wrong mode of operation. At that time the engineer has the opportunity to remedy the source of the error by demonstrating proper use of the equipment. If an operator continues to use the equipment in an improper fashion, permanent damage could occur to the equipment which may require an expensive repair.

For clarity, the four most common types of service calls related to radiographic equipment are listed below in order of frequency of occurrence:

1. ***Mechanical failures*** as a result of equipment handling (and mishandling), worn parts, or missing hardware.
2. ***Electronic failures*** as a result of components being subjected to heat or power surges and from repeated cable stress.
3. ***Image quality problems*** as a result of some failure in the imaging chain.
4. ***Operator error*** as a result of improper inservicing of the staff or the addition of new personnel who are unfamiliar with the equipment.

Most basic mechanical repairs can usually be performed by entry level service engineers (i.e., level I). However, intermediate level engineers (level II) are needed to troubleshoot and repair most electronic problems. Level II engineers can also resolve some minor imaging problems. All imaging problems and the most difficult electronic failures can best be handled by senior level engineers (level III).

This chapter will cover different troubleshooting techniques that have proven successful for the author, followed by a review the common failures found in radiographic rooms. By using the troubleshooting procedure described in this chapter, the service engineer can *quickly diagnose* these common x-ray failures and perform repairs properly so that no *service "recalls"* will be necessary.

Proper Documentation

The experienced service engineer should be able to narrow down the cause of an x-ray failure on most types of equipment without referring to the service documentation, the reason being that all x-ray systems are very similar in their basic functions.

However, once the problem has been isolated to a specific area, then a good, updated, service manual is required to troubleshoot "down" to the actual component that is causing the failure. Although all brands of x-ray equipment contain circuits which perform similar tasks, the operation of these circuits varies greatly among manufacturers. A good service manual, complete with schematics, a theory of operation, and a complete parts list, is a valuable tool necessary for troubleshooting radiographic equipment.

When to Call for Help

When a service person arrives at a site with equipment in a "down" situation, the obvious goal is to get the customer operational as quickly as possible. If the engineer follows the troubleshooting sequences described below, the cause of the failure will be identified within a relatively short period of time, perhaps as short as one hour. The engineer can then attempt a repair or, he or she may have to order parts and return at a later date for the repair.

If, however, ***two hours*** have passed and the service engineer still has no clue as to the cause, or even the area of the failure, then it is probably time to call for assistance. Many times, phone assistance from a technical support person is all that is needed to get the engineer on site pointed in the right direction. If phone support does not help to resolve the problem, then a second service engineer with a fresh approach to the problem should be dispatched to the site.

It is often difficult for an engineer to admit that he or she is unable to resolve an x-ray problem, especially in situations where egos are involved. Everyone likes to think that he or she is infallible and can solve any problem encountered. This is a foolish attitude that will only cause problems for the customer, as well as the service organization. *No one can fix every x-ray problem that may arise!* In fact, a second opinion often results in a much shorter service call. Being able to call for help without hesitation or reservation, and knowing the right time to call for help are qualities that any respected x-ray service engineer must have.

TROUBLESHOOTING TECHNIQUES

Generally speaking, it is difficult to teach the student troubleshooting skills in a classroom setting. Many of these skills are acquired through trial and error while working in the field. Clearly, nothing can replace actual experience as the best teacher of troubleshooting techniques. The goal here is to provide a general method for troubleshooting common x-ray failures that has proven successful for the author. Following this method, a service engineer should be able to resolve most x-ray problems in a reasonably short period of time.

Each individual service engineer will have his or her own method of successfully attacking a service problem, and that method may vary from the procedures given below. This is expected and is totally acceptable. There is always more than one way to skin a cat!

The first task of the service engineer, when he or she arrives at the site, is to find out what exactly occurred before and after the equipment failure. When enough information regarding the failure is obtained, the engineer must then try to observe the failure and reproduce it, if possible. This step is most difficult when intermittent failures occur. Once the failure can be reproduced, the engineer will then isolate the area of the failure through a specific sequence of actions. The troubleshooting sequence must be performed in a timely manner so that any repairs can begin as early into the service call as possible. Once the repairs are completed, the system must be fully tested to ensure that the failure has been corrected.

Gathering Information on the Failure

The service engineer must first gain information as to when and how the failure occurred. Like any investigator, he or she must gather information regarding the incident (i.e., failure) from the witnesses who were present at the time the mishap occurred. Was the equipment operating properly in days prior to the failure? At what time of day did the failure occur? Was the equipment cold or had it been "on" for a period of time? In what mode of operation was the machine being used at the time of the failure? In what positions were the table, tube support, and x-ray tube when the failure occurred? Did the failure occur with just one operator or did several people observe the same failure? Was the radiologist present? Did any unusual sounds or odors accompany the failure? These are just some of the important questions that the service engineer should ask when he or she arrives at the site.

Also, the engineer should find out when the unit was last serviced and what most recent repair was made on the unit. Oftentimes the answers to these questions will give important insight into the current problem. In fact, the service call may be related to, or even have been caused by a prior service call. Service engineers are not exempt from making errors, and for this reason the most recent repair should be considered when diagnosing problems.

While the service engineer is gathering information about the failure, he or she must be aware that the person providing the description is recalling from memory and, consequently, may leave out (or add) important details. Since the x-ray technician is busy attending to the patient's needs, he or she is not really concentrating on the equipment, per se. In addition, most x-ray "techs" are not technically oriented (it is not a job requirement) and usually do not have an in-depth understanding of how the radiographic equipment really operates.

Since the operator's account of the failure might possibly be inaccurate, the service engineer must decide whether or not the information can be relied upon. It is common to hear of situations where the engineer was led (for hours!) in the wrong direction because the technician did not give an accurate account of the failure. For this reason, if the service engineer regularly visits the particular site, he or she should identify a person who is reliable at describing technical details and should ask that person to relate the details of the failure. If at an unfamiliar site, the service engineer must use his or her own judgment as to the reliability of the description given.

Another common scenario that all service engineers will experience at least once in their career is the situation where the engineer will ask the staff to try performing a few quick checks on the equipment, or the engineer may ask if a specific mode of operation was checked. If the staff confirms that they have already done this, the engineer, in an attempt to save time, will not bother to test that particular function and will move on to other areas to find the cause of failure. Eventually, hours (or days!) later when the cause of the problem has still not been found, the engineer will decide to check the area that was previously bypassed.

If it turns out that the problem was, in fact, in this area, the engineer will have learned a valuable lesson that will not soon be forgotten. The best choice is for the service engineer to personally test all suspected areas and not to rely on a staff member to do this.

Observing the Failure

Once the service engineer has gathered enough information regarding the failure, the next task is to try to ***observe*** (i.e., reproduce) the exact failure, noting the sequence of events that led to it. If a component is obviously damaged, this step is simplified and the engineer should then begin the repair phase of the service call. With ***intermittent failures,*** however, the engineer must take certain steps to ensure that the original failure will be reproduced.

The most important thing to remember when attempting to reproduce an intermittent failure is *not to alter anything in the room* until the failure can be reproduced. Many service engineers will walk into a room and manipulate a suspected cable or reseat a connector, or they will clean a switch contact that, from experience, he or she feels could be the cause of the problem. This is precisely *the wrong thing to do!*

If the conditions in which the room failed are changed, and the failure cannot be reproduced while the service engineer is at the site, it could falsely be concluded that he or she has "fixed" the problem. In reality, there is no way of knowing if the actions taken by the engineer have indeed "fixed" the problem. Service engineers who use this unreliable troubleshooting technique are not doing their job properly and waste a good deal of their time on return service calls (i.e., recalls) for the same problem.

The correct troubleshooting approach is to operate the equipment as the x-ray technician normally would and see if the failure can be reproduced. In fact, *the best method is to let the x-ray technician demonstrate the failure to the service engineer,* if possible. By doing so, the service person can see whether the operator is doing something incorrectly to cause the failure or if the equipment is actually malfunctioning.

This troubleshooting technique will greatly aid the engineer in diagnosing x-ray failures. Once the failure can be reproduced, a major part of the troubleshooting phase is over. The service engineer can then perform the necessary steps needed to isolate the cause of the problem and then begin to take correct action.

It is easy to see why this phase of troubleshooting is considered by the author as the most difficult when we consider the occurrence of

highly intermittent failures. Failures that are highly intermittent are the most serious (and frustrating!) type of failure that can occur in any mechanical or electronic device. Moreover, these failures seem to occur at the worst possible time when the equipment is most needed.

This is especially true for medical imaging equipment since when a radiographic room intermittently "goes down," many people are directly affected. The patients who had appointments for that day must be called and rescheduled for another time. Those patients that have already arrived at the office must be sent home, which often creates problems, especially if they have made special arrangements for transportation.

In addition, when the x-ray unit is "down" the radiologist not only experiences a significant loss in revenue since he or she cannot examine patients, but also must pay the salaries of staff members even though they are not actually working. Everyone must simply wait for the service engineer to arrive. This frustrating situation is especially devastating to a medical facility that has only one x-ray room on which to rely.

Because of the untimely nature of these failures, it is apparent that intermittent problems can be a major source of aggravation for both the staff and the service engineer. Furthermore, it often happens that when the service person arrives at the site, the equipment will function properly. If the service engineer cannot observe the equipment failure, he or she then must regretfully inform the radiologist that the problem could not be reproduced and, therefore, no corrective action was taken, and that the failure most likely will occur again. For this reason it is vital to follow a specific troubleshooting sequence to try to reproduce the intermittent failure.

Finding the Highly Intermittent Failure

If the equipment is functioning properly upon his or her arrival at the site, the service engineer will often leave the site immediately because there is nothing else that he or she can do. This is another incorrect approach when troubleshooting intermittent problems. By utilizing the information concerning the failure given by the staff, the service engineer should begin checking all of the suspected areas.

The engineer should first begin to examine in detail all suspected areas for the usual signs of intermittent problems. These signs may include loose connections, worn cable coverings, frayed wires, discolored components, or excessive heat on individual components. If no obvious signs are found, the service engineer should begin gently pulling and tugging on individual cables in the suspected areas to see if the failure can be re-created. ***Intermittent cable connections*** are a likely cause of intermittent failures. If there is an intermittent connection involved, moving the cables around will exaggerate the connection problem and, therefore, cause the failure to occur.

If this technique of ***forcing the failure*** does not produce the problem, then each device in the suspected area should be mechanically repositioned and then further tested. This includes raising (or lowering) and tilting x-ray tables, moving the x-ray tube through its entire range of motion, and repositioning wall stands. By repositioning the suspected device, the equipment cables may be flexed into a position that will exaggerate the intermittent connection. It is often the case that the equipment will only fail in a certain position.

If no failure occurs after testing the suspected device in different positions, the service engineer should disconnect a suspected cable and inspect the individual pins for corrosion or loose crimp connections. The individual wires of the cables can also be tested for continuity. If everything appears in order, it can be concluded that the failure is not related to a faulty cable. The engineer must then consider if the intermittent failure is occurring on a circuit board.

When probing around circuit boards, components should be examined for damage and solder joints should also be visually inspected for sound connections. In addition, the individual components should be gently "rocked" in order to find poor connections that cannot normally be seen. It is often helpful to use a magnifying glass when examining connections on circuit boards.

If the cause of the failure is not yet apparent, the engineer should note if the failure occurred when the equipment was first turned on and "cold." If the failure occurred at that time, the equipment should be turned off and allowed to cool. When the components are sufficiently cool, the unit should be turned back on and the troubleshooting sequence should be continued. If the failure occurred after the unit had been operating for a period of time, the failure could be heat-related and troubleshooting should, therefore, be performed when the unit has reached normal operating temperature.

It must be noted that with highly intermittent failures, it may take some time before the failure reappears. A partially broken (stranded) wire or a loose connection may never show up while the service engineer is on the site and may remain functional for weeks or months before it will finally fail again. Also, the connection might conceivably fail only one time while the engineer is troubleshooting and then will function properly for the rest of the time the engineer is at the site. For these reasons, the service engineer must be persistent in his or her search for the source of the problem.

If the intermittent connection does fail, the engineer must be alert to all events that took place prior to and during the failure. Most good service engineers are very observant and will notice an unusual sound or action that accompanied the failure. A good approach is to monitor several key test points in the circuit in question so that when the failure occurs, the service engineer will be able to isolate the problem to a specific area.

If, after the engineer has performed the troubleshooting sequence described above, the failure has not yet been observed, it is time for a different course of action. When several hours have passed without an occurrence of the failure, the engineer should conclude that the "highly" intermittent failure will not be found–at least on that particular day.

With highly intermittent failures, logic and common sense rules do not always apply. The failure could be caused by some defect in the materials used in a mechanical device or an electrical component. This type of defect may never be discovered. Consequently, the engineer must use a new strategy as a last resort. In short: ***change a component in the circuit that is suspected as causing the failure.*** The service engineer must make some (obvious) change in a circuit or device that is in question. The engineer may decide to change a questionable relay, transistor, or a complete circuit board. He or she is, in effect, making an educated "guess" as to the likely cause of the problem.

By making a distinct change in the circuit, two important goals are accomplished. First of all, the engineer, by pure luck, may have fixed the problem–the intermittent failure never occurring again. This often happens and is why, although a last resort, this troubleshooting technique is very useful. Secondly, if the failure reoccurs, the service engineer has eliminated one likely component and can troubleshoot with slightly better odds.

When an engineer gets to the point of "blindly" replacing every component in the circuit in hopes to resolve the problem, the repair will become unjustifiably expensive. This desperate troubleshooting

Figure 36. Block diagram of typical x-ray circuit.

technique, called ***shotgunning,*** should not be used by an x-ray service engineer.

Isolating the Cause of the Failure

If the service engineer is able to reproduce the failure, the next step in troubleshooting is to isolate the actual problem area. This is usually done by dividing the circuit in question into two distinct sections or "blocks" and then isolating the failure to one of the blocks. Figure 36 shows a typical block diagram of an x-ray circuit.

With most electronic devices, a command is sent from the control section of the device which, in turn, causes a specific action to occur. The control signal can be a simple command sent from a switch or relay, or a more elaborate series of commands sent from a microprocessor board. For example, on tilting radiographic tables, the tilt motor must receive a drive *command* before the motor will actually begin tilting the table. In another example, the x-ray generator must receive a command from the exposure control circuitry before the exposure will be initiated.

The diagram above also shows a ***system interlock*** block located midway between the control block and the action lock. As a rule, most x-ray circuits utilize ***safety interlocks*** to ensure that the device will be operated without risk of danger to the equipment, the patient, or the staff. The interlock may be located in the control logic, or physically located in the device itself. X-ray tubes utilize ***thermal interlocks,*** for example, that will inhibit an exposure if the temperature of the tube housing becomes excessive. In addition, ***rotor interlocks,*** present in all rotor controllers, will block the exposure if the anode is not rotating. The x-ray generator has ***exposure interlocks*** that will not allow an exposure to occur if certain conditions are not met: such as whether or not the door of the radiographic room is closed; whether or not a film is present in the receptor; or if the tube is positioned correctly. Finally, motor driven x-ray tables have safety interlocks (i.e., collision sensors) that will not allow the motor to drive if something is in the path of the table.

Technically speaking, most safety interlocks consist of a set of relay contacts that are wired in series with the control signal path, so that in the event that a safety condition is not met, the command signal path will "open" and, as a result, there will be no possibility of the signal reaching the intended device. Since the interlocks effectively disable key functions in radiographic equipment, this is an obvious place to begin troubleshooting. On a service call, *the service engineer should first check that all safety interlocks are functioning correctly and are not, in fact, causing the failure!*

If an interlock is not the cause of the failure, the engineer must decide if the failure is caused by a problem in the control circuitry or by a problem in the action circuits. In the words of many experienced service engineers, "Is the device being told to operate or not?" To do this, the engineer should initiate a specific command and then check to see if that command is reaching the appropriate device. In the case of the tilting table, a good approach for the service engineer would be to activate the tilt switch while monitoring the voltage at the tilt motor and note if the command is actually getting there. With this approach, the area of the failure can be isolated within a few minutes.

If the signal is, in fact, measured at the motor, then a major part of that circuit has been eliminated and the engineer can begin testing the motor drive section. If, however, the voltage was not present at the motor, the engineer must then find the point in the control signal path where the signal is being interrupted. In this case, several approaches to troubleshooting work equally well. The service engineer can begin measuring at the source of the control signal (i.e., where the command signal originated) and then sequentially follow the signal path until the signal is lost.

Another approach utilizes the "divide and conquer" tactic. With this approach, a test point midway in the circuit between the motor and the "source" is checked to see if the signal is present there. If present, one half of the circuit has been quickly eliminated. The next step is to find a half way point between the new test point and the motor. This process continues until the faulty component is found. Again, either of these approaches works well and generally takes about the same amount of time.

The above troubleshooting procedure is rather straightforward and can be used for most ***simple action devices*** found in a radiographic room, such as devices that have motors or electromagnetic locks. The x-ray generator, however, is more complicated in design and requires multiple input signals before it will respond with the proper action–the production of x-ray radiation. To produce x-ray radiation, a filament voltage must be applied to the tube, the correct primary voltage must be present, and all the safety interlocks of the generator must be satisfied. These conditions must be satisfied before an exposure command will be "released" from the exposure control logic. Once the exposure logic is satisfied, an exposure command is sent to the timer control circuitry which, in turn, signals the start of the exposure.

If the x-ray generator utilizes digital control, the inputs to the exposure control logic must be checked to see if they are present. The service engineer must then check to see if the exposure logic is outputting the correct signal to initiate the exposure. Troubleshooting these complex circuits requires an in-depth knowledge of x-ray generators and is, therefore, usually performed with the help of an experienced service engineer.

There are other instances of equipment failure in a radiographic room that require an experienced engineer. For example, there are times when several failures will simultaneously occur in different circuits, producing the same symptom. With this type of ***multiple failure,*** one cause that is contributing to the failure may not reveal itself until the other failure has first been corrected. In effect, one failure "masks" the second failure. Troubleshooting and isolating this type of compound failure is understandably more difficult and can be very frustrating to the service engineer. Although multiple failures are very time consuming, the engineer feels a great deal of satisfaction when he or she can solve these more complex x-ray problems.

Verifying the Cause of the Failure

When the device or component suspected of causing the failure has been found, an important task must first be performed before the engineer can be begin making any repairs. He or she must first confirm that the particular component is, in fact, the absolute cause of the failure. There should not be any doubt, whatsoever, in the service engineers mind that the repairing (or replacing) of that particular component will, in fact, completely resolve the problem.

To achieve this high level of confidence, the engineer must try to ***simulate the function*** of the component suspected of causing the failure by using a few common electronic troubleshooting "tricks." Here, the engineer will electronically "mimic" the exact function of the faulty component by the use of these troubleshooting "tricks" to see if the circuit will function properly. If normal operation returns to the circuit, the engineer can be confident that the suspected component really is bad. This troubleshooting technique works well for most active components such as switches, relays, transistors, and logic gates.

One trick commonly used in electronics is to use a jumper wire (with clip leads) to bypass the suspected component completely. By placing the jumper wire across the connections of the faulty component (i.e., in parallel), the engineer is "fooling" the circuit so that the component appears to be outputting the desired signal. This technique works well for switches, relay contacts, and any interlocks that are in question. If the device begins to function properly with the jumper in place, the faulty component is confirmed, with a high degree of confidence, as bad.

As demonstrated above, a simple jumper wire can be valuable (and inexpensive) troubleshooting tool for the service engineer. Another "trick" used by experienced engineers is to artificially inject the correct signal at the input or output of the suspected component. This technique makes the suspected component appear to be functioning so that the rest of the circuit can be tested. Service engineers successfully use this troubleshooting technique when testing digital logic circuits.

For example, suppose a logic gate must output a "high" signal level (e.g., +5 VDC or +15 VDC) to enable further circuits to function. Instead, however, the gate outputs a "low" signal and, therefore, is suspect. The engineer, in this instance, can physically inject a high level signal at the output of the gate. If the circuit begins to function properly, the suspected component is confirmed as bad and, in addition, the engineer will know that the rest of the circuit is functioning. When using this technique, service engineers say that they are ***forcing the test point high.*** Conversely, injecting a "low" signal is referred to as ***forcing it low.***

To force an input or output to change states, the service engineer can again use the handy jumper wire. To force an output "high," the engineer will clip one side of the test jumper to a +5 V (or +15 V) supply and then touch the other side of the test jumper to the output pin of the component in question. On many circuit boards, a +5 VDC (or +15 VDC) test point is conveniently provided for testing purposes. The same technique is used to force a low logic state, only the jumper lead is now connected to the 0 volt point or a ground point.

This technique of injecting signals *should only be performed by an experienced service person with a strong electronics background.* Because of the high cost of circuit boards and the sensitivity of the components used on the boards, great care should be taken when injecting signals. Injecting a voltage at the wrong point could further damage other components in the circuit. In addition, there may be multiple problems on a circuit board that will corrupt the result obtained by using this troubleshooting technique. For example, a second component on the board could be damaged, causing the output of the component in question to appear not to be functioning correctly. Again, only an experienced service engineer will be able to resolve compound problems of this sort.

The true test that confirms that the cause of the failure has been found is when the service engineer actually replaces the component. With the new component in place, the engineer should observe a drastic change in the operation of the circuit. If the circuit behaves in a similar manner with the new component installed, this indicates that the wrong component was replaced and the engineer should continue troubleshooting that circuit. *The engineer must observe the operation of the circuit before and after the repair to actually confirm that he or she has successfully diagnosed the problem.*

Performing Repairs

When the failure has been correctly identified and the replacement parts, if needed, are "in hand," the engineer can begin to repair the equipment. As a rule, the repair should be the easiest part of the service call. An engineer possessing good electronic skills, a fair mechanical ability, and plain common sense should be able to perform most repairs. After the work is completed, further testing will confirm that the repair was performed properly.

All service engineers should know how to perform repairs at the component level. Here, a good soldering/desoldering technique is imperative in order for the repair to be effective. Oftentimes, the replacement component could become damaged if heated for a period of time by the soldering iron. In addition, the ***heat*** from the soldering iron can easily damage lands on circuit boards or melt the insulation of wires that are in the area of the repair. Since this is counterproductive to the repair, the engineer should always be aware of the amount of heat that is being applied to the component or circuit. In addition, heat sinks and IC sockets should be used when replacing sensitive electronic components.

There are occasions when a service engineer must ***modify*** a circuit during a repair in order to get the device operational. A ***field modification*** is a physical alteration of the original circuit that is performed on site. Generally, the engineer will have to modify a circuit if he or she cannot obtain the part(s) needed to repair the device. The engineer may also modify a circuit that he or she decides is not needed for the particular installation, or may disable a circuit that "never really worked properly."

In addition, if the cost of a replacement part is exorbitant, the radiologist may decide not to have the device repaired and will simply not use that particular x-ray function anymore. In this case, the engineer must entirely bypass the circuit in question.

Consequently, service engineers do perform field modifications on a regular basis. There is one caveat that should be mentioned in regard to equipment modifications. For many important reasons, *most manufacturers of radiographic equipment must be notified before any alterations can be performed on their equipment.* As a rule, any unapproved field modifications performed on new equipment will immediately void the manufacturers warranty on that equipment. Furthermore, if a patient or staff member gets injured while using the modified equipment, the manufacturer will not be liable for damages and the service engineer could be sued for performing the modification without written approval.

The service engineer must be aware of the consequences that could result from a field modification. Once the engineer has received written approval for a field modification (for new equipment) or has decided that a modification can be performed without a reduction in performance (or safety!), he or she can then perform the modification.

When the modification is complete, the engineer must make a record of exactly what was done to the device.

If the engineer is modifying a circuit, he or she must *document the modification in the service literature.* Proper documentation includes a full description of the alteration with accompanying detailed drawings. By documenting the circuit modification, the engineer is responsibly performing his or her job so that anyone who attempts to service this equipment in the future will be aware that the equipment has, in fact, been altered.

Without the proper documentation, a service engineer (other than the one who performed the "mod") will not be able to perform troubleshooting properly since the original schematics are no longer correct. Many times, an engineer will troubleshoot a circuit for many hours because it does not seem to be working properly, only to later find out that it was modified and is working the way it was intended. The extremely frustrated service engineer must then start troubleshooting the original problem over again.

Equipment Testing

When the repair has been completed, *the service engineer must thoroughly test the functions of the x-ray system.* Final testing should not be limited solely to the device that was repaired since most individual devices in the room are interconnected to form an x-ray system. Testing should verify the operation of the device that was repaired, as well as the system interlocks that may be related to the functions of the device. For example, if a bucky motor drive circuit was repaired, the engineer should test the operation of the bucky using x-ray exposures to ensure that the exposure interlocks are functioning correctly.

The greatest amount of service "recalls" result from the service engineer not thoroughly testing the system after making a repair. The reason for testing all functions of the x-ray system after a repair, including those functions that have absolutely no relation to the circuit that was repaired, is partly the result of the compact design of today's radiographic equipment. The fact that the internal components are so closely positioned to each other often makes repairs difficult to perform.

Usually, a wire harness must be moved or disconnected to reveal the circuit that is to be repaired. In fact, with some repairs, a major assembly may have to be completely removed from the equipment cabinet before the failed component can be accessed. Furthermore, when working in tight spaces, the engineer could inadvertently disrupt an unrelated circuit without his or her knowledge.

This information should not surprise the reader of this book. Problems caused by the engineer (i.e., "self-induced" problems) do regularly occur in the x-ray servicing field, as is the case with all other service professions. However, if the engineer checks every function in the x-ray system before he or she leaves the site, most of these secondary problems will be discovered and immediately corrected. Although a functional checkout may take several minutes to complete, many engineers fail to perform these checks. As a result, these same engineers also perform a lot of service recalls.

Troubleshooting Sequence

To recap, the troubleshooting sequence that the service engineer must follow to successfully resolve x-ray failures is as follows:

1. Gather as much information as possible about the failure.
2. Reproduce the failure.
3. Troubleshoot to component level.
4. Verify that the suspected component is actually bad.
5. Check that the new component is functioning properly.
6. Test all functions of the x-ray system.

COMMON FAILURES WITH RADIOGRAPHIC EQUIPMENT

There are many different ways in which a radiographic device could fail. Any attempt at describing the numerous failures that could possibly occur with radiographic equipment would fill many long and tedious volumes and simply would not be practical. However, since a major goal of this book is to prepare the student who will soon begin a career in this field, some specific examples of equipment failures will be described in the following pages.

The "generic" failures that are described below commonly occur among all brands of radiographic equipment and every service engineer will at some time encounter most of these failures during his or her career. This information, combined with the x-ray and imaging fundamentals provided throughout this book, will enable the service engineer to confidently troubleshoot and repair any failure that can occur with radiographic equipment.

Failures with the X-ray Control

The x-ray control regulates the production of x-ray radiation. It also controls many other functions in a radiographic room. For example, the tube safety circuits, all x-ray interlocks, rotor control, and AEC are functions handled by the x-ray control. It is "the brain" of the x-ray generator. Consequently, failures can occur in the control that could affect many other devices in the system. The discussion here, however, will be limited to specific failures with the output radiation that result from a malfunction in the x-ray control.

The most devastating failure of the control is when there is no radiation being produced, and so we will begin our discussion here. In order to produce x-ray radiation, three components are needed (assuming a functional x-ray tube): ***high voltage,*** or ***kVp; tube current,*** or ***mA;*** and an ***exposure start signal.*** If one of these signals is missing, x-ray production will not occur. We will first examine kVp failures.

Common KVp Failures

All x-ray generators use an *autotransformer* for kVp control. The "auto" takes the incoming line voltage supplied by the power company and provides the multiple output voltage "taps" that are used to vary the ***primary voltage.*** The operator will select a particular kVp setting on the console which, in turn, will activate a particular tap on the "auto." The selected tap will send the appropriate primary voltage to the high voltage transformer produce the desired kVp.

As mentioned in Chapter III, a special switch, called a ***primary contactor,*** is installed in series with the primary circuit. The primary contactor will close at the start of an exposure to supply the selected primary voltage to the primary winding of the high voltage transformer. A typical contactor is made of two "back to back" SCR's (silicone controlled rectifier), although some older x-ray generators may still use a large relay "contactor" or thyratrons for primary switching.

The primary contactor is switched on by an exposure start command. The exposure start command originates from the operator handswitch, goes through the control logic and the timer circuitry, and finally to the primary contactor. Though very basic in design, the primary circuit will experience a few common failures from years of use.

Since as much as 300 amps (momentary) can pass through the components of the primary circuit, it is reasonable to assume that over time they will experience failures. When a failure occurs in this circuit, several high voltage problems will result.

When troubleshooting, the engineer should examine the active components in the primary circuit. These include: the ***line contactor;*** the ***safety contactor;*** the ***kVp major*** and ***kVp minor mechanical tap switches;*** and the ***primary contactor circuit.*** In addition, the integrity of all connections in the primary circuit must be verified by the engineer. A loose primary wire will create an intermittently resistive connection that will directly affect the primary voltage and, hence, x-ray output.

The ***line contactor*** is energized when the x-ray unit is turned on, and remains energized throughout the day until the unit is powered off. Consequently, it will show signs of wear after some years of use. Specifically, the contacts used in line contactors will become burned and pitted from being subjected to the extremely high currents of the primary circuit. Damaged contacts will become intermittently resistive and will result in fluctuations in the primary voltage. If the primary voltage fluctuates, the kVp will likewise fluctuate.

The service engineer should routinely inspect these contacts during a PM. If the contacts are found to have minor damage, they can easily be repaired. The engineer can carefully polish, or ***burnish*** and clean the contacts to restore the contact surfaces. If the contacts are severely pitted, they should be replaced so as to avoid further problems.

The same type of contact failure can occur with the ***safety contactor*** and the mechanical ***kVp tap switches.*** A safety contactor is present in most x-ray controls and is installed directly in series with the primary contactor. The safety contactor will open the primary circuit if a failure in the safety circuits is detected. Failure conditions such as a tube ***mA overload***, a technique that ***exceeds the generator rating,*** or a ***filament failure*** will activate the safety contactor and, as a result, will inhibit the production of x-ray. The safety contactor is subject to the exact same currents as the primary contactor and, therefore, fails in the same manner.

With ***kVp tap switches*** that are commonly used in single-phase generators, the internal contacts will also become burned and pitted over time. Usually, the failure will first appear on the most commonly used kVp station only. If the engineer monitors the output of these switches (i.e., the primary voltage), he or she will notice a significant drop in primary voltage when the tap is selected–or there will be no voltage present at all. When tap switches fail, they cannot be repaired in the field and, therefore, must be replaced.

Another common failure in the primary circuit occurs with the ***primary SCR contactors.*** The SCR circuit can fail in a number of ways to cause different kVp problems. If one of the two "back to back" SCR's becomes shorted, high voltage will be present at the tube instantaneously when the safety contactor closes. Since the exposure time cannot be controlled with a shorted SCR, radiation will be emitted from the x-ray tube until the safety contactor "drops out."

This is a serious condition that could cause the patient to receive an excessive amount of radiation. To remedy this problem, many manufacturers incorporate SCR failure detection circuitry in their x-ray generators. In these circuits, the exposure logic constantly monitors the SCR's for a shorted condition. If an SCR breaks down, the logic will inhibit the exposure.

If an SCR "opens," only one-half of each AC cycle will reach the primary of the high tension transformer. This failure results in a condition that is termed ***half-waving.*** If the primary of the HVT consecutively receives only the positive (or negative) alternations of the AC cycle (i.e., one half sine wave), excessive current will flow through the transformer causing the line breaker to trip.

The excessive current is caused by a phenomenon that occurs in every solid core transformer, called ***core saturation.*** The core of a transformer, usually made of iron, normally should see consecutive pulses that are opposite in polarity–a positive pulse followed by a negative pulse, for example–to function properly. If the transformer receives two consecutive pulses of the same polarity, the core will become saturated and the transformer will not function properly. The sole function of the ***core memory circuits,*** located in the timer circuitry, is to ensure that each exposure begins with the correct polarity (see Chapter I).

The reason why the core becomes saturated has to do with the eddy currents that are flowing though the core of the transformer. When the core is saturated, it becomes highly conductive, thus causing excessive currents to flow through the core, instead of through the secondary windings. The high current flowing through the core will severely load the power line causing erratic exposure to occur. Most of the time, however, the severe loading will simply trip the main line circuit breaker.

A telltale sign of a transformer that is experiencing core saturation is a very loud humming sound that will be heard coming from the transformer. Half-waving, however, is easily identified by monitoring the primary voltage at the transformer with an oscilloscope. Either the positive or negative alternation of the AC cycle will be missing on the waveform.

Another failure with the primary contactor circuit occurs in the ***gate control circuitry*** of the SCR's. The gate pulses are sent from the timer control to turn on the SCR's to start the exposure. *The SCR's must be signaled "on" by the time control logic to start the exposure.* Consequently, a failure in the gate control circuit will cause several x-ray failures.

A complete failure of the gate control circuitry, where no timing pulses are produced, will result in a no exposure condition. This failure is easily diagnosed with an oscilloscope placed at the output of the gate pulse circuit. Another gate control failure that is often more difficult to diagnose occurs when the gate pulses are present, but are at the wrong amplitude or pulse width, or incorrectly timed. This becomes even more critical with three-phase and high frequency units.

Common MA Failures

MA control is accomplished by providing a closely regulated voltage supply to the primary windings of the filament transformer, which is located in the high voltage transformer unit. The ***filament stabilizer transformer,*** located in the x-ray control cabinet, provides this regulated supply.

When an mA station is selected, a specific voltage is sent from the "stabilizer," through the mA adjusting resistors, to the primary of the filament transformer. During the exposure, the filament secondary current will be in the range of 4.0–5.5 amps, which represents a significant load to the supply. The stabilizer transformer is used to ensure that the filament voltage and, therefore, filament current remains constant during the exposure. Maintaining this regulated output puts a great demand on the stabilizer. Consequently, filament stabilizer transformers can fail over a period of time and must be replaced.

Time Control Failures

The time control circuitry performs several functions. It initiates the exposure with an ***exposure start command,*** it controls the exposure time, and it monitors the line alternations for the core memory circuits. As a rule, the timer board in any generator has a very low failure rate and is rarely repaired. Other than the failure of the core memory circuits that causes the half-waving phenomena described above, the timer could fail but this would be a rare occurrence.

Failures with the High Voltage Transformer Unit

The main function of the high voltage transformer unit is to "step up" the selected incoming primary voltage to a high voltage which will then be ***rectified*** (converted to DC) and applied to the x-ray tube. A second important function is to step down the selected primary filament voltage to a level that can be applied to the filaments of the x-ray tube. Thus, the major components inside the high voltage transformer unit are the ***high voltage*** (step up) ***transformer,*** the ***rectifier diodes,*** and the ***filament*** (step down) ***transformer.***

Each of these components has the potential to fail during the life of the x-ray equipment. Although there are only a few components involved, failures here are often difficult to diagnose and repair. Furthermore, because of the potentially dangerous voltages that are present in the transformer unit, the engineer is severely limited in his or her troubleshooting techniques. To complicate matters even further, the components in the transformer unit are completely submerged in insulating oil inside the sealed tank. Consequently, it is not a common practice to service the internal components of the transformer unit on a routine service call.

The engineer must, therefore, rely on measurements made externally to diagnose problems in this area. The key to solving high voltage problems is having the ability to recognize the different symptoms of typical transformer unit failures. Also, the engineer must quickly rule out any effects of the x-ray tube or high voltage cables in order to isolate the transformer unit as the cause of the failure (see following sections).

The two components in the transformer unit that are directly involved with the production of high voltage are the high voltage transformer (HVT) and the stick rectifier diodes. Since the high voltage transformer, itself, rarely fails, it would be the last component to suspect when diagnosing a high voltage failure. The "stick" rectifier diodes do, however, commonly fail in two ways.

The "stick" rectifier is made of several high voltage diodes soldered together in series on a circuit board that is similar in appearance to a common yardstick. Four of these "sticks" comprise a full wave bridge rectifier circuit in single phase generators. When they fail, a stick rectifier will either "open" or become "shorted."

A shorted stick rectifier is easily detected because of the severe effects it will produce. Any exposure that is attempted with a shorted stick rectifier will produce a very loud "bang" originating from within the transformer. In fact, the entire transformer unit may actually vibrate if the exposure technique is high enough. The reason for the severe reaction is that a shorted stick will cause a direct short across the secondary of the HVT during one of the alternations of the AC primary voltage. Once observed, and only one exposure attempt is needed, the service engineer can cease troubleshooting and can place the order for a new set of sticks (the x-ray tube and cables should have been eliminated as the cause at this point). The entire set of rectifier diodes should be replaced since other diodes could have been damaged during the failure.

If the diodes in the stick have "opened," a completely different symptom will be observed. In this case, only one-half of the AC cycle will reach the x-ray tube. By ***half-waving*** the x-ray tube, only one-half of the selected mA will be attained. There will be no noticeable sound or action to indicate an opened stick. The radiologist will place a service call complaining of underexposed films when using a manual exposure technique.

The service engineer can easily identify an opened stick by two methods. The simplest way is for the engineer to observe the mA panel meter on the x-ray control. If the unit was known to be calibrated, a reading of exactly one-half the set mA could indicate an opened stick. To confirm this finding, the engineer should view the high voltage waveform to verify half-waving.

Before concluding that a stick is actually opened, the engineer should make one more check. The failure could be related to a problem with the primary circuit in the x-ray control (see X-ray Control Failures). If the primary is correct, then the engineer can confidently order the new stick rectifiers.

The other major component in the transformer unit that can fail is the ***filament transformer.*** The filament transformer should output a voltage in the range of 8.0 to 12.0 VDC, depending on the type of tube used and the mA selected at the control. The tube rating chart that was supplied with the x-ray tube will list these voltages. More importantly, the secondary current of the transformer must be in the range of 4.0 to 5.5 amps, depending again on the particular mA setting, to produce radiation.

If the filaments are not "glowing," the engineer should first check that the primary filament voltage is present. Convenient test points, located on the top of the high voltage transformer unit, are provided for making these measurements: ***XL*** for large focus, ***XS*** for small focus, and ***XC*** for filament common. If the primary voltage appears correct, the engineer can then insert test equipment in the secondary circuit and verify that the correct voltages are being outputted.

To make these measurements, the engineer must pull the cathode cable from the well of the x-ray tube. He or she must then attach a test lamp to simulate filament loading.[27] A voltage meter is place across the test lamp, and a current meter (AC clamp) is connected around one of the leads to the lamp (Figure 37). The engineer can now initiate prep (first handswitch position only!) and make accurate measurements of filament voltage and current. If the output is not correct, the high voltage cable may be at fault (see below) or the filament transformer is actually bad.

This test setup should also be used when replacing an x-ray tube that has a "blown" (i.e., opened) filament. The reason for doing this is that the opened filament in the original tube may have been caused by excessive current (above 6.0 amps) in the filament secondary. If the

27. The test lamp should have characteristics that are similar to the actual filaments in the x-ray tube. In doing this, the engineer can obtain realistic measurements. As a rule, this circuit should always be tested under load.

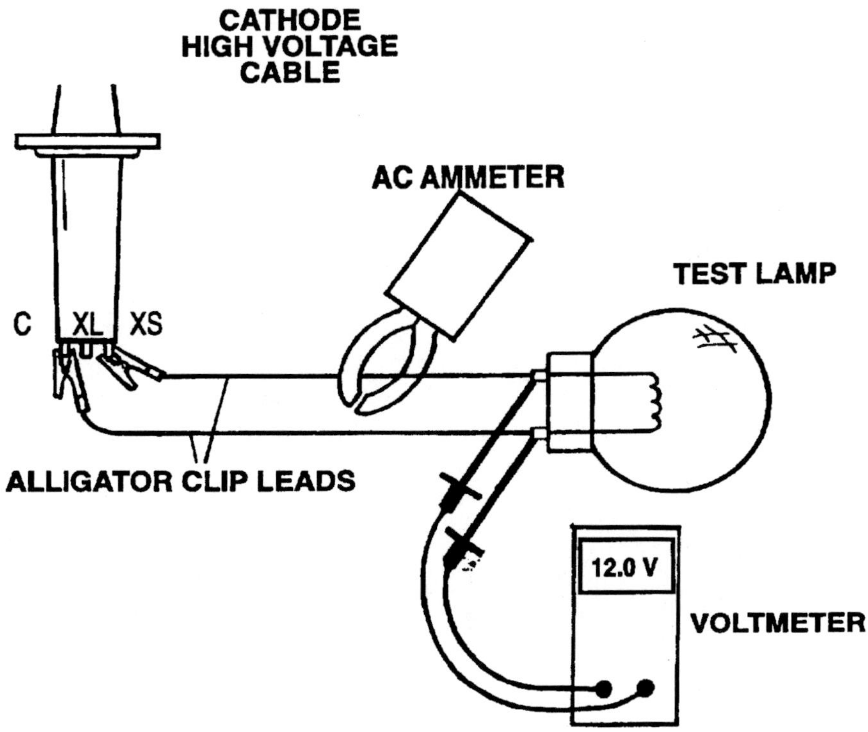

Figure 37. Filament test setup. This circuit is used to verify that both the filament voltage and current are correct.

new tube is installed without first making this check, the filaments in the new tube could be permanently damaged.

Failures with the X-ray Tube

X-ray tubes commonly fail in a number of ways. For example, a filament could open, high voltage arcing could occur in the tube housing or within the tube insert itself, the rotating anode could fail, the radiation output of the tube could drastically fall off, the resolution (i.e., focus) of tube could change, the anode could become pitted causing erratic exposures, the housing seal or the glass insert could break causing an oil leak, or a receptacle could get damaged. Since x-ray tubes are very expensive (costing roughly $5,000.00 and up), it is imperative that the service engineer exercise great care when installing or handling them. The engineer should also ensure that the tubes are operating within the specified safety limits so that they will provide trouble-free operation for many years.

Damage Caused by Handling

The engineer should demonstrate great care whenever handling x-ray tubes. To ensure that no damage will occur, x-ray tubes should be repackaged in their original shipping carton when being transported to the site. Careless handling of a tube will usually cause the glass insert to crack which renders the tube useless. When installing a tube, the engineer should make sure that the tube is safely mounted (using a thread locker, such as Locktite) and that all connections at the tube are carefully secured. If, for example, the rotor wires (or terminal posts) of the x-ray tube are damaged during the installation, the tube must be sent back to the reloading facility to be repaired.

Furthermore, care should be taken when threading the cable nuts from the high voltage cables into the tube wells. If a cable nut is accidentally cross-threaded into the threads of the x-ray tube well, the threads could be permanently damaged and the tube will have to be sent back to the reloading facility for repair.

Heat Related Failures

If an x-ray tube is properly installed and regularly calibrated, the engineer can expect approximately 10 years of tube life in a radiographic setting. This extended tube life is in part due to improved tube manufacturing practices and the fact that exposure times are much shorter for most examinations performed today. The shorter exposure times are a result of using faster film/screen combinations and also from the use of high frequency generators.

It is a fact that *most failures with x-ray tubes are a direct result of damage caused by heat.* As heat accumulates within a tube, it begins to cause internal damage that will accelerate the aging of the x-ray tube. By taking shorter exposures, the technician is effectively reducing the amount of heat to the tube and, therefore, increasing tube life.

The anode bearings of rotating anode x-ray tube are extremely susceptible to the effects of heat. Over time, the heat will cause a breakdown in the lubricant of the internal rotor bearings and, as a result, the anode will begin to produce a distinct audible sound during rotation.

This sound will become increasingly louder over time and, at some point the bearings will "cease up" and the anode will no longer rotate.

Excessive heat could also damage the tube housing. A rubber seal, called the ***bellows,*** is specifically installed in the cathode end of all x-ray tubes to allow for oil expansion caused by heat. Because of size limitations, the bellows can only compensate so much for heat expansion. Consequently, if too much heat is accumulated in the tube, the seal could burst.

Anode Failures

Since the target material of the anode is constantly bombarded with high speed electrons, it will eventually begin to show signs of wear. If the x-ray tube was installed correctly, is regularly calibrated, and the operator strictly follows the daily ***tube "warm up" procedure,*** the target should remain in good condition or many years. Problems arise only when the above criterion are not followed.

If a cold target (i.e., from a tube that was not used for several hours) is hit with a high technique exposure, the thermal stress caused by the contrasting temperatures on the target surface could cause the target to become warped, and no longer useable. Thermal stress can also cause pits or cracks to form on the target surface. If the target is pitted (or cracked), erratic exposures will most often result.

Most anode problems, however, are a direct result of ***anode rotation problems.*** The anode of a rotating anode x-ray tube must rotate at the correct speed (and direction) so that the maximum exposure ratings of the tube can safely be achieved. In fact, the exposure rating of a tube with a rotating anode is directly restricted by the speed of anode rotation. During an exposure, the stream of electrons that impact on the anode produces a great amount of heat (recall that 99% of the energy used for x-ray production is dissipated as heat). The anode is rotated specifically to dissipate the heat evenly across the entire anode surface.

If the anode fails to rotate, the x-ray tube will be permanently damaged. Usually, the area of impact on the target surface of a ceased anode will actually melt and become permanently deformed. Even if the engineer repairs the anode rotation problem, the tube will remain unusable. The reason being that each time the electrons impact the damaged area of the target, the resulting radiation will be unpredictable and erratic, causing inconsistent exposures.

Anode rotation failures can occur if the rotor bearings cease (from heat damage) or if there is a failure in the rotor controller. The engineer can quickly narrow down the failure by removing the end cap of the x-ray tube (anode end) and attaching the leads of an oscilloscope to the stator shift and stator common leads. A second probe should be connected the to the stator main lead.

An AC voltage should be present at the tube during the boost and run cycle. The boost voltage for low speed rotation should be about 200 VAC and the run voltage should be in the 60 VAC range. In addition, the stator main and shift voltages should be out of phase when viewed on the oscilloscope. The degree of phase shift is dependent on the shift capacitor used on the rotor controller and the type of x-ray tube used.

If the voltages are correct, then the anode bearing assembly probably has ceased and, consequently, the tube must be replaced. If no voltage is present, or an incorrect voltage is present, the problem is with the rotor controller. Most low speed rotor controllers are located inside the x-ray control unit. It basically consists of a power supply, a relay, and a timing circuit to control the delay time between the boost cycle and the run cycle. The engineer should check these components to find the cause of the failure.

If the anode is rotating, the rotation speed (in rpm's) should be measured with a vibrating reed tachometer (see Chapter VI). An anode that is rotating at a speed that is much slower than its rated speed will also fail prematurely. In this case, the anode will accumulate excessive heat that it will not be able to dissipate properly. Over time, the anode will become warped or pitted causing erratic exposures, and the tube will have to be replaced.

Filament Failures

In addition to anode failures, the cathode of the x-ray tube will also experience a few failures. These failures will mostly occur with the filaments of the x-ray tube. Usually a filament wire will either become "open" or it will become "shorted." During an exposure, the filaments are heated and, in turn, will emit electrons. Along with the emission of electrons, a small amount of the tungsten material is also released in a process called filament evaporation. The evaporated tungsten will be deposited on the inside wall of the glass insert. Excessive tungsten deposits could lead to arcing within the tube insert (see below).

Over a period of years, filament evaporation will cause the filament wires to gradually become "thinner" and very brittle. Just as with the collimator lamp, when a very brittle filament is subjected to a high inrush current, it will "break" open.

In the case of an open filament, the radiologist will usually report that he or she is not getting an exposure on either large or small focus. Usually, the service engineer can easily confirm that a filament has, in fact, "opened" by removing the collimator and looking directly into the port of the x-ray tube, first with the generator powered off. It should be absolutely obvious that the filament wire is broken. This method of direct observation is the most reliable method for diagnosing an opened filament wire.

If, however, the tungsten deposits on the port of the tube are partially obstructing the view of the filaments, the engineer can measure the resistance of the wires. The resistance measurement should be made at the pins in the cathode well of the x-ray tube.[28] Again, the bad filament should be obvious, as indicated by an extremely high resistance reading. If the engineer obtains similar readings between large and small focus, the problem may be elsewhere (see high voltage cable failures).

The engineer can also inspect the filaments with the generator powered on. With power turned on, the engineer should alternately select between large and small focus and confirm that one filament is not lighting. The engineer should then "prep" the x-ray generator in order to observe the filaments under load. In this case two engineers are necessary, one operating the x-ray handswitch and one viewing into the port of the tube. Most importantly, *the service engineer must view the filaments, through an optical filter to protect his or her eyes from the intense light that emanates from the filaments!* If a filter is not available, a piece of exposed and developed x-ray film can be used as a filter.

Caution: the tube must only be "prepped" (i.e., first position on handswitch). If an exposure is accidentally taken during this test, the engineer in the room will be exposed to dangerous amounts of radiation!

Another way the filament will fail is that it can short to the focusing cup of the cathode. With a "shorted" filament, the radiologist will expe-

28. Since it is difficult to reach these connections, a test connector can be made from an old spare high voltage cable. The cable is cut six inches back from the federal connector and the three conductors should be stripped back to expose the bare wire. For testing, the engineer will insert the test connector in the well of the tube (with power turned off!) now using the bare conductors of the cable for the resistance measurements.

rience a generator overload condition when attempting to make an exposure. When the generator detects an overload condition, the control circuits will automatically shut down the generator completely. The overload is a result of excessive tube current caused by the shorted filament. The shorted filament will draw excessive filament current which, in turn, creates a very large space charge that produces the high tube current.

A "shorted" filament is also relatively easy to confirm. Many times the engineer can look directly into the port of the x-ray tube during standby (i.e., no radiation) and see the shorted filament "sagging" and actually touching the focusing cup of the x-ray tube. Another way to confirm a shorted filament is by making resistance measurements.

If the filament is, in fact, shorted, the engineer must find the cause of this failure. It may be that the x-ray tube is very old and the filament failed as a result of the normal aging process. If the failure was caused by excessive currents in the filament circuitry, the engineer must correct this problem before installing a new tube.

Low Radiation Output

An x-ray tube failure, although very rare, relates to the radiation output of the tube and is caused by a problem with the filaments. The engineer will discover this failure during a routine calibration, or on a service call for a complaint of light films. Here, the desired mA cannot be achieved, regardless of the amount of filament current present. The low mA output results in a low radiation output. This type of failure mostly occurs with tubes that are very old, or in tubes that have been heavily used. It is rare to find this failure in a new tube.

The failure is very straightforward. To produce a specific tube current (mA) at a given kVp, the filament is heated to a certain level by applying the required filament voltage. In fact, the optimum voltage (and current) required to produce a given mA can be read directly from the ***filament emission chart*** for that particular tube. If the actual mA value cannot be achieved at the specified filament voltage, or even if a higher voltage is used, the tube is probably bad and must be replaced.

This rare type of filament failure is caused by excessive or extended use of an x-ray tube. A sufficient amount of the filament evaporation has occurred to the point where the filament can no longer liberate enough electrons to produce the higher mA outputs. Usually, the tube will fail in another way before it gets to this point.

X-ray Tube Arcing

Internal arcing is another common way in which an x-ray tube can fail. A high voltage arc can occur by two means: (1) an alternate path of conduction within the glass insert and (2) a breakdown or reduction of insulating oil within the tube housing. In either case, the arcing that occurs within the tube housing is potentially dangerous to persons who may be in the vicinity of the x-ray tube.

Furthermore, the high voltage arcs will often cause additional damage to other circuits in the x-ray system. Usually, the arc will travel back through the system causing damage to some other component in the x-ray generator. The numerous, highly sensitive gates, proms, and solid state devices used in x-ray generators are most often damaged by high voltage spikes.[29] This is why it is crucial to diagnose high voltage arcing failures as quickly as possible.

High voltage arcing will most often occur within the glass insert of the x-ray tube. As stated earlier in this chapter, tungsten deposits will eventually accumulate along the inside walls of the glass insert. Overtime, excessive deposits will create a conductive path for the beam of electrons to follow. Since the beam of electrons normally travels through the evacuated space of the x-ray tube (which is highly resistive), a path of tungsten metal will provide a less resistive path for the electrons.

The tungsten, in effect, is providing a direct short from the anode to the cathode of the x-ray tube. The short circuit will cause excessive tube current which creates on overload condition. Most tube overloads will shut down the generator completely.

Internal tube arcs are very easily detected and can be diagnosed in a matter of minutes. The first obvious sign is the loud "snap" or "pop" that is heard originating from the x-ray tube. Once the engineer confirms that the sound is emanating from the tube, he or she must then remove the collimator and aim the port of the tube towards the "lead shielded" control booth. With the lights turned off, the engineer will then take an exposure while observing (from behind the lead glass window) the port area of the tube. A bright light should be seen inside the

29. This is why the engineer must wear a grounding wrist band and use an antistatic pad when changing IC's. A static electric discharge is, in effect, a high voltage spike, as demonstrated by the accompanying audible "snap" that is heard.

tube that coincides with the loud snap. If confirmed, the tube must be replaced. A sign that a tube has been arcing is by the presence of ***crazing*** along the port of the x-ray tube. Crazing has an appearance of fine cracks along the plastic window of the port.

Another type of high voltage arc occurs within the tube housing assembly. In this case, the insulating oil may have degraded in quality from being exposed to heat or it may possibly contain impurities. Most of the time, however, a small amount of oil has leaked from the tube causing an air gap to form within the housing. In some cases, the housing was never completely filled at the reloading facility.

Regardless of the method of failure, the high voltage arcs that occur within the tube housing can cause as much damage to the x-ray system as arcing originating from within the glass insert. Although the path of conduction is outside the glass insert, the arcing produces the same results (i.e., it will damage components or simply confuse the memory circuits of the x-ray generator causing a "shut down"). On the positive side, the tube does not have to be replaced. With housing arcs, the tube can be repaired at the tube reloading facility for a relatively low cost and then reinstalled.

Arcing can also occur within the high voltage receptacles at the x-ray tube or high voltage transformer. This failure results from the absence or breakdown of the vapor proofing compound (at x-ray tube) or transformer oil (at transformer) and is easily identified by the presence of carbon tracks in the well. Most of the time, the engineer can correct this problem by cleaning the well and reapplying the grease or oil. On the rare occasions, the well will be permanently damaged and must be replaced.

"Gassy" Tubes

Another type of failure that an x-ray tube may experience is related to gas that may be present within the glass insert. In theory, as the filament material becomes "vaporized," a small amount of gas will be released. This gas is normally absorbed into the inside surface of the glass wall of the tube insert. If, however, the tube is not used for a period of time, or if it has been stored on a shelf for several months, the surface of the glass may begin to release these gas molecules.

Any gas molecules in the path of the electron beam will create a conductive path that will, in effect, cause an internal short within the glass insert. As with any internal arcing, the excessive current will cause the x-ray generator to shut down. With tubes that have been inactive for very short periods of time, say several months, this problem could be remedied by seasoning or conditioning the tube (see Chapter VI).

However, after a much longer period of inactivity (or if there is a vacuum leak), an excessive amount of gas will accumulate within the glass insert x-ray tube. In this case, tube seasoning will not correct the situation. The large number of gas molecules will not be reabsorbed into the glass surface and the tube is said to become ***gassy.*** A gassy tube must be replaced.

Failures with the High Voltage Cables

The high voltage cables connect the x-ray tube to the high voltage transformer. Although it would seem that very little could go wrong with a pair of interconnecting cables, a failure with these special cables can cause a service engineer to completely misdiagnose a problem in a radiographic room.

The most common problem occurring with high voltage cables is that one or more of the conductors within will eventually break "open" after years of routine positioning of the x-ray tube. These cables can absorb a fair amount of stress and, over time, the constant stress will cause the wires to fray and eventually break. For this reason, the importance of good cable draping must be again emphasized.

Although any interconnecting cable can experience stress and broken wires, a broken conductor in a high voltage cable will produce unusual symptoms that can confuse the service engineer when he or she is diagnosing a problem. In a small signal cable, a break will represent an "open" circuit and the attached device will simply not function. With high voltage cables, the high voltage signal will tend to "jump" across the air gap formed by the break in the wire.

A cable arc will often produce symptoms similar to an x-ray tube arc and can, therefore, mislead the service engineer. The arcing could cause the x-ray generator to shut down or it may cause an erratic kVp output. With software controlled generators, the high voltage spiking will damage other sensitive components in the system and cause a system lockup. In this case the x-ray generator power must be cycled off and on to reset the computer logic.

If an opened conductor is suspected, the engineer must try to observe the arc during an exposure. This must be done with the room lights completely "off" since the arcing occurs within the heavy rubber insulation of the cables and is not easily observed. If the arc is seen, the engineer must replace the cable since they are not repaired in the field.

As a rule, it is always best to test high voltage cables "in circuit," as they are normally used. If the engineer so decides to pull the cables to make resistance checks, or to "charge up" the cable using a high voltage bleeder, he or she may disturb the conductor which is broken and could actually obtain results that would indicate a good cable.

The conductors in the high voltage cable could also, from stress, become shorted to each other. This is not a problem for the anode cable since the three conductors are shorted together at the high voltage receptacle anyway. It is, however, a major problem for the cathode cable, which uses the three conductors for the filament circuit as well as for conducting high voltage. A shorted cathode cable will cause filament problems and could cause damage to the filament transformer. Since it is very costly and time consuming to replace a high voltage cable, there is a quick and easy remedy for a shorted cathode cable. In this situation the service engineer will simply swap the anode and cathode cables at both the x-ray tube end and high voltage transformer end ***(both ends must be swapped!).*** Again, the conductors of the anode cable are physically shorted anyway.

Broken or frayed conductors in a high voltage cable can also mislead the engineer when diagnosing a filament failure. As said, the high voltage cathode cables also supply filament current to the filaments of the x-ray tube. The actual secondary filament current is usually in the range of 4.0 to 5.0 amps, which represents a significant amount of current. When the conductor of the high voltage cables are overflexed, the individual strands of wire in that conductor will begin to fray. As more strands are broken, the conductor will become increasingly resistive.

Finally, a point will be reached where there are only a few strands of wire conducting the large filament current. A voltage measurement may indicate that the cable integrity is good, even if the strands of wire are broken and simply resting against each other. However, when tested under load, a significant "IR" drop will occur at this resistive point. Since the filament wires themselves represent a very small resistance, most of the filament voltage is dropped across the resistive connection and as a result, the filament will not light up.

Failures with the Collimator

As stated in Chapter IV, the main function of the collimator is to limit the size of the radiation field. It accomplishes this beam limitation by the use of mechanical shutters. The blades of the shutter assembly open and close to vary the radiation field size. Manual shutters, because of their simple design, generally provide trouble free operation during the functional life of the collimator-barring any major damage caused by collisions, for example. Automatic collimators have motor driven shutter blades that are also very reliable. The most common failures that occur with collimators relates to the ***lamp circuit failures*** and ***PBL system failures.***

Failures in the Lamp Circuit

The most common collimator failure occurs with the ***field lamp circuit.*** Since the purpose of the field lamp is to visibly indicate the radiation field, the field lamp has to be bright enough to be seen in normal room lighting. To achieve this brightness (minimum of 160 lux), a power supply capable of supplying a relatively high current (as much as 6 amps with most collimators), a means of switching the current on and off, and a special collimator lamp are required.

The high current of the lamp circuit has a two-fold effect. First of all, it will most often damage the components in the lamp circuit used for "switching" the lamp "on." The two most common devices used for switching in collimator lamp circuits are relays and triacs. When switched "on," the initial high in-rush current (usually in excess of 10 amps) can damage these components as they age. For this reason, it is common for a service engineer to replace these components in the lamp circuit.

In addition, because of the initial surge of current, the collimator lamp itself will occasionally burn out. With age, the filament in the lamp becomes brittle due to filament evaporation and cannot handle the high inrush current. Consequently, the filament will instantly pop at "turn on" (Recall that when a household lamp "blows," it is always when it is being switched on).

Besides the shock caused by excessive inrush current, the high current in the lamp circuit also generates a great amount of heat within the collimator housing. The collimator lamp, itself, gives off a tremendous amount of heat when producing the high intensity light required to project a visible field. Unless the lamp is fitted with the proper heat shielding, the heat from the lamp will cause damage to the collimator circuitry over time.

PBL Failures

The other circuit in a collimator that most often fails is the automatic (PBL) circuit. Most PBL failures are a result of the collimator shutters not moving to their correct position, or not moving at all, when the cassette is inserted in the cassette tray. Invariably, this failure is caused by an ***incorrect or missing collimator drive signal.*** This signal originates at the cassette size sensing circuitry of the cassette tray, and is sent to the collimator logic to cause the shutter motors to "drive." The service engineer can easily find the failure by following the troubleshooting sequence below:

1. Check that the cassette size sensing tray is outputting the correct resistances for each cassette size. (The correct resistances are listed in the Chapter VI).
2. Check that the correct voltages (directly related to the resistances) are reaching the collimator control.
3. Check that the shutter motors are getting the correct drive signal.

Some collimators have a special feature that allows for the convenient testing of the collimator logic. By placing the collimator in a "test mode," signals can be injected into the collimator logic to simulate the insertion of a particular size cassette. This test mode is used to eliminate the collimator head as the cause of the failure. It 9 out of 10 cases, the cassette tray is the cause of a PBL failure.

Failures with the Receptor

The receptor in a radiographic room can be either a basic grid cabinet or a more sophisticated bucky assembly. Both types of receptors utilize a cassette tray to hold the film cassette in place. Because of the constant loading and reloading of film cassettes, the cassette trays are subject to mechanical failures.

Usually, the cassette tray will exhibit many failures during the useful life of the equipment. The rubber grip assemblies that hold the cassette in place will wear down over time and the locking knob that holds the cassette in place will also fail with excessive use. In addition, the rear connector of the tray, used for PBL operation, is a common area of failure. These connectors will get damaged if the cassette becomes misaligned, or the contact pins may simply wear down with use. Regardless of the type of failure, the cassette tray connectors are replaced regularly in radiographic rooms.

If a bucky assembly is installed, there are some electromechanical failures that will occur. The reciprocating grid assembly is usually the source of most bucky problems. A symptom that often indicates a grid failure is a complaint of grid lines on the film. The grid lines are caused by the fact that the grid is no longer moving during the exposure. It should be recalled that a reciprocating grid has a lower grid ratio and grid frequency (Chapter VI), resulting in much thicker lead strips than those of stationary grids.

The thicker lead strips will be clearly visible on the film if the grid is not moving, or if moving at the incorrect speed. With a reciprocating grid failure, the grid itself could be jammed, the grid motor could have failed, or there could be a problem with the linkage (i.e., gears, or belts) between the motor and grid assembly.

Failures with the Tube Support, Wall Receptor, and Radiographic Table

During an x-ray examination in a general radiographic room, the tube support, wall receptor, and radiographic table are constantly repositioned for each patient that is examined. Because of this repeated movement, these three devices are prone to mechanical failures. Common ***mechanical failures*** occur when parts loosen as a result of the vibrations and shocks caused by the movement. In addition, many mechanical parts are damaged as a result of equipment abuse (i.e., collisions, overtightening of locks, and other careless handling).

Besides mechanical failures, the most common area of failure for tube supports and radiographic tables is with the electromagnetic locks. The ***electromagnetic locks*** on radiographic equipment will exhibit failures during the life of the equipment for three reasons. First of all, the power supplies for most electromagnetic locking systems are generally of very low quality. Manufacturers try to save expense in this area and will use the most basic power supply designs, often made from lower grade components.

Electromagnetic lock supplies are designed to operate at the minimum level of performance needed to ensure that the locks will "hold." A typical DC lock power supply might contain a 24 VAC transformer and two diodes that are used for rectification. This unfiltered, unregulated, full wave rectified lock supply is prone to failure. One telltale sign of a failing lock power supply is an annoying buzzing sound heard when the locks are energized.

Another reason for the high rate of failures has to do with the fact that the electromagnetic locks in a radiographic room are usually operating in the ***energized state*** for most of the working day.[30] The locks are de-energized only when the tube, table, or wall unit is being repositioned while examining a patient. While the locks are energized and holding the device in place, electric current is flowing steadily through the locks. This constant current heats the locks to the point that they can often be hot "to the touch." This heat, applied over a period of time, will cause damage to the coil within the lock, or the wire insulation of the lock.

In addition, electromagnetic locks become magnetized when current is flowing through them. The magnetic attraction of the lock is what causes it to move toward the metal "holding" surface of the particular device. Since the lock moves towards the locking surface, the wires attaching to the lock also move. This repeated movement could stress the wire leads of the lock, especially if it is rotated, and cause them to break.

The electromechanical lock works by friction. As a result, the material on the surface of the lock will slowly wear away over time. This condition is exaggerated when technicians attempt to move the device without first releasing the locks. In addition, if the locks are not aligned squarely with the locking surface, excessive wear and even damage could occur. For this reason, it is important to properly adjust all locks at installation and then check the adjustments at every PM.

Another component on a table or tubestand that commonly fails is the ***centering and travel limit switches,*** and their associated detents.

---

30. Electromagnetic locks should not be confused with the solenoid style lock. Solenoid locks normally remain in a deenergized state and are activated when repositioning the tube support or table.

Centering switches, as the name implies, are used to indicate the center alignment position in a radiographic system. The x-ray technician must be sure that the receptor is exactly centered to the x-ray tube before "taking" a film. At the time of installation, micro switches, and their detents, are permanently installed so that they will activate a signal when the tube is properly centered to the receptor. In addition, travel limit switches are commonly used in motorized tables, tube supports, and wall receptors.

The micro switches used in these circuits are often generic in design, utilizing lightweight actuators. Here is another area where equipment manufacturers attempt to save production costs. These generic switches cannot withstand the excessive use of the mechanical devices in a radiographic room and, consequently, will break over time.

Image Quality Problems

As stated in Chapter VIII, troubleshooting imaging problems can be difficult and must be performed by an experienced service engineer. The engineer must have a complete understanding of the effects that each device in a radiographic room has on the quality of the image. The information provided in Chapter VIII will enable the service engineer to effectively troubleshoot image quality problems. Here we will discuss the most common causes of image quality problems. When these causes have been ruled out, the service engineer should begin to evaluate the complete imaging chain as described in Chapter VIII.

Some of the most common complaints concerning image quality are: films that are too dark or too light, complaints of grid lines on the image, an unsharp or "blurry" image, an image with too low or too high contrast, blank (or totally black) films, and artifacts on the film. Before the engineer should attempt troubleshooting the x-ray system, *he or she must first rule out film and film processing as the cause of the imaging problem.*

All of the image quality complaints listed above could be attributed to a problem with the film or film processor. Light films could be caused by bad or old chemistry in the film processor. The engineer can quickly check the condition of the developer by removing the processor covers and confirming that the color of the developer is light yellow in color. Developer that appears brown in color is considered bad and must be replaced. Another test to check the condition the developer is to dip a film that has been exposed to light directly into the developer for a minute to see if it will turn black. The fixer can be tested similarly by dipping an undeveloped film into the fixer tank. The film should be completely clear when taken out of the tank.

Also, if the developer temperature is too high, darker films will result. The service engineer should always carry a thermometer on service calls. Light or dark films can also result from the wrong film accidentally being used. In addition, the film could be expired. A quick way to check out both the processor and film is to expose a film (in a cassette) with the film ID camera that is used to put the patients name on the film. If the name is dark and clear after processing, the chemistry is probably good and the film correct.

A complaint of grid lines could turn out, instead, to be roller marks from the processor. Artifacts seen on the film could also be caused by water marks or scratches that were caused by processing. The service engineer should inspect the processor to see that it is generally clean and that the rollers are in good condition, and that the water filters have been changed.

Film artifacts could also be caused by static electricity or dirty screens in the film cassette. Static electricity has always been a source of problems in radiography. In the winter months, when the building is heated continuously, the relative humidity often drops well below the range specified for x-ray film (i.e., 30%–50% RH). Consequently, the extremely dry environment is conducive to a buildup of static charges between the different materials (i.e., plastic and rubber) used in cassettes and film loader/unloader units. Artifacts caused by static electricity are easily identified by their distinct pattern they leave on the film: wavy lines emanating in all directions from a point.

A blurry image could be caused by ***poor film/screen contact*** in the film cassette. The engineer can use a test wire mesh, available in most x-ray departments, to check proper film/screen contact. Poor image contrast could be caused by ***film fogging.*** Light originating from a light leak in the darkroom, film cassette, or a cracked (or the wrong type) safelight filter could "fog" the x-ray film. The engineer can test for film fogging by developing an unexposed film and then measuring the gross fog with the densitometer.

Every factor related to film and film processing that could have an effect on image quality should be methodically ruled out before the engineer begins diagnosing the radiographic equipment. Most of these checks can be performed very easily, in a matter of minutes. It would be futile to waste time attempting to replicate a failure that never occurred, or to break down a system to install invasive equipment to view perfectly good waveforms.

It is common to hear of cases where the engineer has spent several hours testing a system, only to later discover that a new lot of film (of the wrong type) had just arrived that day. This waste of valuable time is completely unnecessary and indicates a lack of experience on the part of the service engineer. The point cannot be stressed enough.

When all other factors have been ruled out, and the service engineer is confident that the problem lies somewhere in the imaging chain of the radiographic room, he or she can then begin troubleshooting the x-ray system to isolate the problem area. A chart showing the factors that effect image quality is reproduced in Figure 38. This chart can be used as a guideline for isolating the cause of an imaging problem when manual x-ray techniques are used.

Many image quality problems can also be caused by the ***automatic exposure control unit*** of the x-ray system. A failure in the AEC circuitry will directly result in various film density problems. The films produced with a malfunctioning AEC unit will be overexposed, underexposed, or inconsistent densities may result. However, before the service engineer attempts to recalibrate an AEC system, he or she should first ask several questions:

1. Was the film processor operating properly?
2. What technique settings were used when the density problem occurred?
3. Was the patient positioned properly for the exposure?

As with all imaging problems, the film and film processor must first be checked for correct operation before the engineer can proceed with testing the x-ray system. The engineer must also note the technique settings that were being used when the poorly exposed films were produced. The proper settings must be selected at the console or imaging problems will result. When using the AEC mode of operation, a ***film/screen type, a density sensitivity setting,*** the correct ***detector*** and ***sensing field,*** and a proper ***backup time*** must be selected before initiating an exposure. If any one of these settings is incorrect, the film density will be drastically affected.

IMAGE DENSITY

TECHNIQUE SELECTION
-kVp
-mA
-Time

FILM TYPE

PROCESSOR CONDITIONS
-Developer Temp
-Developer Strength
-Emersion Time

IMAGE CONTRAST

TECHNIQUE SELECTION
-kVp

BEAM QUALITY

GRIDS

PATIENT

FILM TYPE

FILM STORAGE
-Light Leaks

IMAGE SHARPNESS

X-RAY TUBE

SID

OID

PATIENT MOTION

INENSIFYING SCREENS

FILM / SCREEN CONTACT

IMAGE NOISE

FILM SPEED

FILM ARTIFACT

PROCESSOR CONDITION

INTENSIFYING SCREENS
Speed and Condition

BEAM QUALITY

Figure 38. Summary of factors affecting image quality.

Most generators allow for the use of two or three different ***film and screen types*** when in AEC mode of operation. The reason for this is that many hospitals use several film types for their examinations. At the time of installation, the engineer should have calibrated each film/screen setting for the film that will be used at that institution. Usually, there are provisions for a fast, medium, and detail film. If the technician selects the incorrect film/screen type for a particular examination, the film will not be properly exposed.

The ***density selector*** allows the technician to make minor adjustments to the density of the final image. Adjustments are made to allow for various patient size and pathology. Most generators provide incremental changes of +/– 25% of normal density. If the density selector is not set prior to the exposure, the film density will be affected.

The ***detector*** that is being used (i.e., wall or table) must be selected at the console before an AEC exposure is attempted. Although it would seem obvious, a great many service calls result from the wrong detector selection. Often the technician will attempt a chest examination on the wall receptor while the table detector is selected at the console. In addition, the correct sensing field must also be selected for the particular examination. If a lateral view of the chest, for example, is desired, the "center" sensing field must be selected. With a lateral chest examination, the two "outer" fields are not covered by the patient. Consequently, if either outer field is selected, an underexposed film will result.

Finally, the appropriate ***backup time*** must be set on the console when using the AEC mode of operation. The backup time is a safety measure necessary in case the AEC fails to stop the exposure. Therefore, it is usually set for time much longer than the expected exposure time. Many times the backup time is set to one second, or more. If the backup time setting is too short, the exposure could be terminated by the generator timer instead of the AEC unit. Usually, if this happens, an indication is provided on the console alerting the x-ray technician that the exposure was, indeed, terminated by the generator.

Patient positioning is the number one cause of density problems related to the AEC mode of operation. The anatomy of the patient to be examined must be positioned directly in front of the "active" detector field so that it will be correctly exposed. If the radiologist wants to view the spinal column, for example, the patient's spine must be positioned directly over the center [active] "cell" of the detector. If the patient's spine is only partially covering the cell, radiation will pass relatively unobstructed through the soft tissue of the patient, resulting in an underexposed film.

Many times, the service engineer can examine the films that were rejected by the radiologist to identify a patient positioning problem. By viewing the films, the engineer will be able to quickly rule out problems caused by patient positioning, and then can proceed investigating other causes for the density problems.

Chapter X

THE MOBILE X-RAY UNIT

Mobile x-ray units, also known as "portables," are commonly used in hospitals throughout the world.[31] Originally designed in 1919, the mobile x-ray unit became popular when it was used successfully during the Second World War in field hospitals. Today, the portable now serves an important function in hospitals by allowing nonambulatory patients to be "x-rayed." The mobile x-ray unit is transported directly to the patient's room where the patient is radiographed.

The mobile x-ray unit is a complete x-ray system on wheels (Figure 39). Included in the unit are the x-ray generator, x-ray tube, tubestand, collimator, and a film storage bin. Some mobile units have an additional motor drive system to facilitate transportation throughout the hospital.

In the design of the portable x-ray unit, several obstacles had to be overcome. First of all, the unit had to be small and light enough so that it could be easily maneuvered down hallways, in elevators, and in patient rooms. Mobility was a problem because the components of a permanently installed x-ray system, in general, are very large and cumbersome. Second, the portable had to produce x-rays using the available power in that particular area of the hospital. This meant that instead of a dedicated power line of the standard of 208 VAC, single phase (or 480 VAC, three phase) used for conventional radiographic machines, the mobile unit had to operate from the 110 VAC, single phase outlets that are accessible from most locations in a medical facility.

31. The term "portable," short for portable x-ray unit, originally referred to a unit that could be carried by hand. Today, most professionals in the x-ray field prefer to use this term over mobile x-ray unit. In fact, a technician taking a radiograph with a mobile unit is said to be "doing a portable."

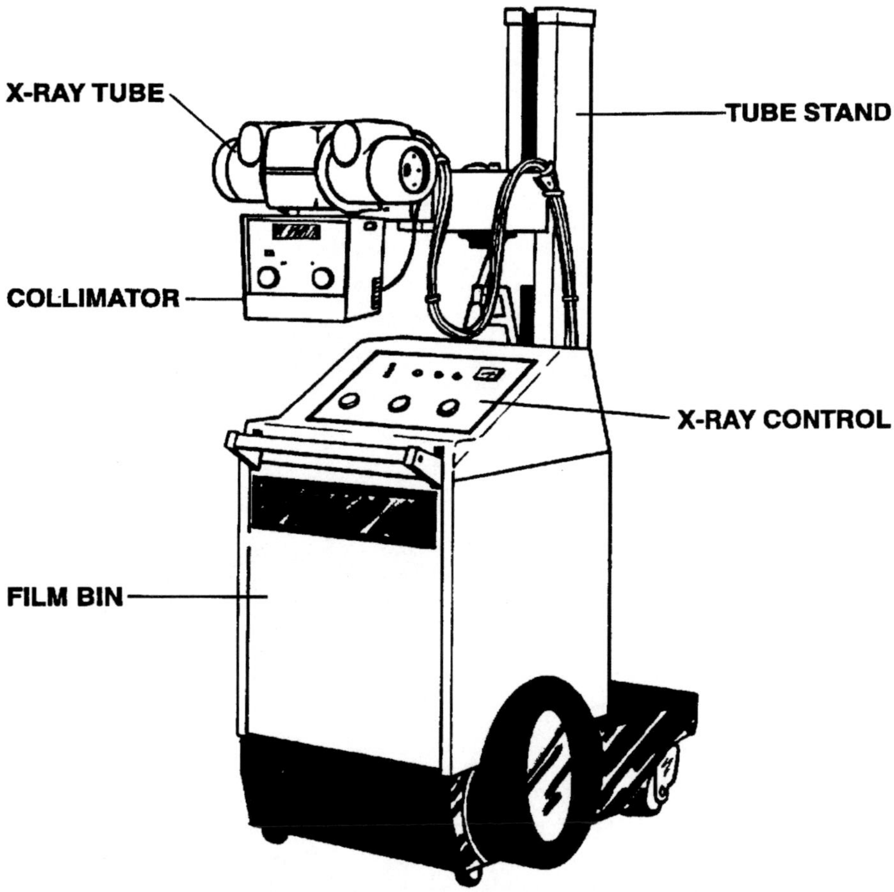

Figure 39. The mobile x-ray unit. A complete x-ray system on wheels.

This second criterion turned out to be the most difficult design problem because all x-ray units require a good quality dedicated power line to handle the high current that is needed to produce radiation. In an attempt to overcome this power problem, three different designs in mobile x-ray units resulted: a 110 VAC ***line-powered*** version, the ***capacitor discharge unit,*** and a ***battery-powered*** version.

LINE-POWERED UNITS

The line-powered version of the mobile x-ray unit is essentially identical in design to x-ray generators used in radiographic rooms. All line-powered mobile units have an x-ray control, a high voltage transformer, and an x-ray tube, but in a smaller version on wheels. Because of the smaller-sized components that are used, and the fact that they are powered by 110 VAC, these portables have a relatively low output level of radiation and, consequently, have very limited use in general radiography. Also, the lower radiation output capability of these units results in much longer exposures that often produce poor radiographs. For this reason, line-powered units are rarely used today except in cases where low techniques are required as in pediatric examinations, for example.

CAPACITOR DISCHARGE UNITS

Capacitor discharge units were specifically designed to address the problem imposed by the limited 110 VAC input voltage. Rather than using direct line voltage for the exposure, these units utilize charged capacitors to deliver the high voltage to the x-ray tube. By using this method, exposures at a relatively high mA can be achieved for a short period of time.

These units operate differently from a conventional x-ray unit. The capacitor discharge unit is plugged into a normal 110 VAC outlet, a technique is selected, and the operator then must press a "charge" button to prepare for an exposure. In fact, an easy way to identify a portable unit as a capacitor discharge type is by the presence of a charge button (not to be confused with a battery charge button). By pressing the charge button, a charging circuit utilizes the 110 VAC line voltage to charge the internal capacitors to a specific kVp. When the capacitors are fully charged, the unit is ready for exposure. *High voltage is now present at the x-ray tube!* The result is that now the capacitors are the power source for the x-ray tube and not the line.

The x-ray tube in a capacitor discharge unit is a special type of ***grid-controlled tube.*** The grid of the tube (i.e., focusing cup) has a bias voltage applied to it when the x-ray unit is turned on. When the unit is charged, the kilovoltage is immediately applied across the x-ray tube. It will not, however, conduct until the grid bias voltage is removed. Depressing the preparation/exposure switch initiates anode rotation and, after a specified time, biases the grid of the tube "on" to initiate the exposure. The capacitors discharge through the x-ray tube during the exposure, producing a negative sloping waveform (Figure 40).

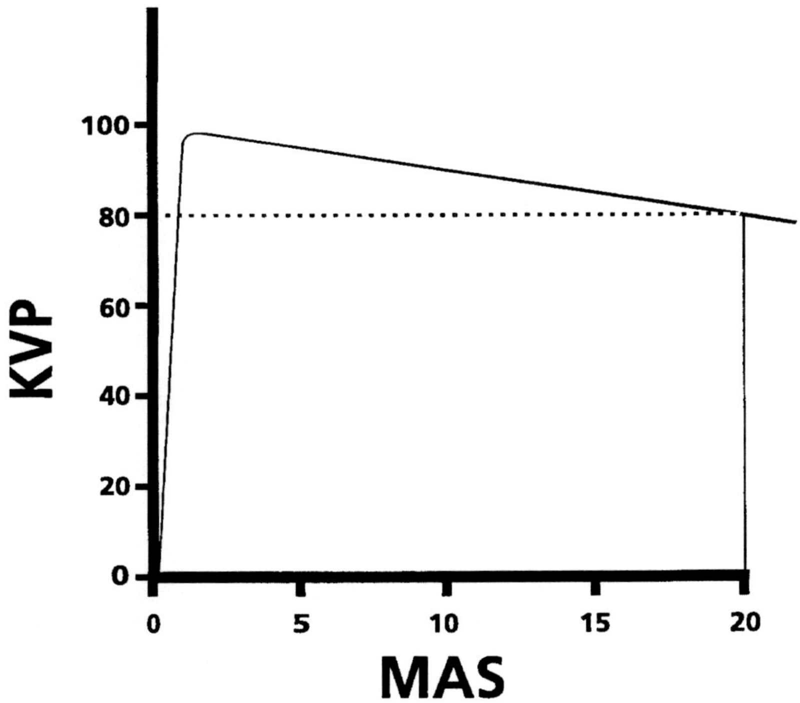

Figure 40. Capacitor discharge unit output waveform. In this typical example, the kVp decreases at a rate of 1 kVp per mAs.

As the capacitors discharge through the x-ray tube, the kVp decreases at a calculated rate (1 kVp per mAs for a 1 μF capacitor), producing a negative sloping waveform. In addition, the actual mA also drops off during the exposure producing the waveform that is similar to the kVp waveform. With the output produced by the capacitors, these units can generate a fairly high technique (e.g., 110 kVp, @ 100 mA) for very short exposures. If the waveform of a capacitor discharge unit is compared to a conventional single phase unit, it will be clear that the capacitor discharge unit produces a much more constant kVp waveform for short exposures.

BATTERY-POWERED UNITS

The third type of mobile x-ray unit provides the best solution to the input power problem. These units are self-powered and do not utilize the line voltage at all for an exposure. The power source for the x-ray generator in this type of mobile unit is a series of high capacity storage batteries. The batteries supply the primary voltage needed to develop high voltage for the exposure.

With battery-powered units, a selected DC voltage, derived from the batteries, is converted to an AC primary signal by means of an ***inverter.*** The AC signal is then sent to the high voltage transformer where it is stepped up to the correct kilovoltage, rectified, and applied to the x-ray tube. Most inverters on mobile units operate at relatively low frequencies (e.g., 500 Hz).

The obvious advantage to this type of mobile unit is that it can be used anywhere in the hospital, regardless of the availability of wall outlets. X ray technicians who operate these units appreciate this feature the most. Other advantages include: an increased radiation output as compared to the other two types of mobile units, a low ripple kVp waveform (due to the higher frequency), and a compact size relative to the level of output these units can produce.

The main disadvantage of battery-powered units is that the batteries must be kept in good condition and properly charged at all times. If the batteries fail to function for any reason, the mobile unit cannot be used. For this reason, a ***battery charge status indicator*** is placed on the front panel of all battery-powered units so that the operator can easily monitor the charge condition of the batteries. This indicator is essential component in the design of battery-powered mobile units because of the characteristically high failure rate of batteries.

Furthermore, a ***battery charging protocol*** must be strictly followed with these units. The battery charging schedule, found in the operation manual, will vary for the type of battery used (i.e., nicad or the preferred lead-acid type). As a rule, the batteries should be charged when the unit is not being used (or overnight). If the batteries are not charged regularly, they will not function to full capacity, will fail often, and will have reduced life.

FAILURES WITH MOBILE X-RAY UNITS

Since a mobile unit is essentially "a radiographic system on wheels," it will experience the exact same kinds of problems that occur with any radiographic equipment. However, because of their specialized design, mobile units will experience a few additional failures. All mobile x-ray units will experience problems with the ***line power cord,*** the ***exposure handswitch,*** and the ***batteries*** used both for producing radiation and to power the drive motors used for transport. In addition, capacitor discharge units will experience unique problems also because of their special design.

The most common failure with mobile x-ray units is a result of a ***damaged AC line power cord.*** AC powered units and capacitor discharge units must be plugged into an outlet (110 VAC) in order to produce radiation. These units are equipped with unusually long power cords (much longer than the standard 15 feet for medical equipment) which are often difficult to store. Consequently, the power cord often becomes tangled and unduly stressed. Moreover, the power cords are left lying on the floor while the mobile unit is being used and, as a result, the cord will invariably get "run over" by various equipment, including the mobile unit itself. Because of the excessive stress, the power cord and plug are regularly damaged.

Also, all battery-powered units must be plugged into an AC outlet when not in use. This is done to keep the internal batteries in a charged condition. The process of plugging and unplugging these units will, over time, eventually damage the pins of the plugs. To make matters worse, technicians often attempt to move the mobile unit, forgetting to first unplug the power cord. This practice will severely damage the plug and line cord. It is easy to understand why this is a high failure area, but it is also easily corrected.

The ***exposure handswitch*** will also fail frequently on mobile x-ray units. As with the power cord, the cable connecting the handswitch to the x-ray control is also unusually long. Consequently, similar damage occurs to the handswitch cable. In addition, the handswitch is usually stored in plastic holder or hung from a hook mounted on the mobile "cart" when not being used. Here, gravity holds the handswitch in place. With this design, the handswitch can be easily knocked out of the holder and, as a result, the handswitch is frequently dropped.

Another common failure in mobile units occurs with the ***internal batteries*** in these units. Most mobile units utilize the lead-acid type storage battery which must be charged whenever they are not in use. In fact, the best way to extend battery life with a lead-acid battery is to keep it "topped off" at all times. Failure to do so will result in poor battery performance and shortened battery life.

Usually someone will inadvertently forget to plug in the mobile unit when he or she is finished using it, or the line cord (or plug) may be broken so that even though the technician plugs the unit in, the batteries will not receive a charge. In either case, this is a detrimental to the battery. If a lead-acid battery is allowed to drain down to 75 percent of the rated terminal voltage, or if left in an uncharged state for a period of time, it will begin to lose the ability to hold a full charge and must then be replaced. Because it is difficult to ensure that the batteries are optimally charged, many manufacturers of mobile units recommend that these batteries be replaced at least once every two years. The batteries should be replaced even if they appear to be functioning properly at that time. This practice greatly reduces mobile unit "down time."

The other failure in mobile units is unique to ***capacitor discharge units.*** With these units, the high voltage capacitors, located in the high voltage tank, store the total kilovoltage charge for the exposure. These capacitors typically will begin to leak current over time and will not be able to hold a complete charge. In addition, the capacitors could become resistive over time. When this occurs, the secondary circuit will be essentially shorted, and the resulting high current draw will cause the line fuse to blow.

Another unique feature of capacitor discharge units is that they use a special ***grid controlled x-ray tube.*** When the charge button is depressed on the control, the unit will charge to the selected kVp which is then directly applied across the x-ray tube. A bias voltage placed on the grid of the x-ray tube will inhibit conduction and, therefore, the production of radiation. When the handswitch is depressed, the bias voltage is removed from the grid via a grid bias switch. The grid switch, because it is located in the high voltage circuit, can fail after a period of use. Also, when normal (3 pin) high voltage cables are used with a grid-controlled tube, the shield of the cable is used as part of the grid switch circuit. Consequently, if the shielding breaks, an exposure cannot be taken. This is a common failure in capacitor units because the cables are regularly stressed during normal use with mobile application.

Chapter XI

TOMOGRAPHY

OVERVIEW

Many of the techniques used in general radiography are effective at obtaining diagnostic quality images of the internal structures of patients. General radiographic techniques are employed when imaging the bones of the extremities or of the lungs, for example, when the desired anatomy is relatively unobstructed. However, certain structures in the human body cannot be directly imaged because they are partially or totally hidden by overlying organs. Also, the desired structure may lie in an area within the body that is surrounded by similar tissue. In these cases, special radiographic techniques are required to obtain images of the obstructed organs.

The organs that comprise the digestive system, for example, are surrounded by tissue of similar density and are very difficult to discern with normal radiographic techniques. Also, some of these organs are hidden by other structures within the body when viewed from an anterior-posterior (AP) projection. One way to view these organs is by the use of ***contrast agents***. When introduced into a structure such as the stomach, a contrast agent will highlight that structure only on a radiograph. Contrast agents are commonly used in fluoroscopic and angiographic procedures to help with the diagnoses of structural anomalies found in the circulatory and digestive systems.

For general radiographic procedures, however, another technique can be used to view structures hidden within the body. It is nearly impossible, for example, to obtain good images of the kidneys by using conventional radiographic techniques because of the presence of overlying organs. In a standard AP projection, the organs of the digestive tract obstruct the view of the kidneys and will be superimposed on the image of the kidneys. Other examples of structures difficult to image in general radiography include the sternum (AP view), thoracic spine (lateral view), and the gall bladder.

The easiest solution when imaging obstructed organs would be to concentrate (or focus) the x-ray beam solely on that particular organ. If the x-ray beam could be focused to the single plane within the body in which the desired structure lies, the resulting image would be of that structure only, regardless of the surrounding organs. This plane specific focusing is accomplished through **tomography**.

Tomography is a specialized application of radiography in which a ***section*** or **"slice"** (tomos is the Greek word for slice) of a patient is imaged. Invented in the early 1920s, body section radiography was perfected in 1929 by Jean Kieffere, a radiologic technologist working in the United States. Over the years, many other terms have been used to describe the technique of sectional radiography, including planigraphy ("plane"), laminagraphy ("layer"), and stratigraphy ("layer"); however, tomography is the term most commonly used today. In effect, tomography allows the radiologist to view a single slice of anatomy at any depth located within the patient's body.

APPLICATION

Once a preferred diagnostic tool for many studies, today conventional tomography has limited uses. With the advent and refinement of CT (computed tomography), many of the procedures once performed using conventional tomography are now being performed on CT scanners. In fact, most of the larger medical institutions regularly employ CT scanners to perform sectional studies. Today conventional tomography is confined to only a few specific types of studies. The primary use for tomography units includes studies of the kidneys, especially ***IVPs (Inter-Venal Pyleography)*** and ***KUBs (Kidney, Ureter, urinary Bladder)*** studies. These procedures are still performed routinely and provide a valuable means to diagnose kidney and urinary functions.

PRINCIPLES OF TOMOGRAPHY

The main goal of tomography is to maintain the precise focusing of a section of anatomy located at one depth within the patient. This goal is accomplished by ***blurring*** the surrounding structures. It is helpful to think of tomography in reverse terms. Rather than "focusing" on a structure lying in a specific plane, we blur the undesirable surrounding structures instead so that only the desired structure will appear focused on the film. In effect, *tomography blurs out the anatomy lying above and below the desired plane, thus emphasizing the specifically targeted structure.* The blurring effect is accomplished by the use of ***motion***.

As with any radiographic procedure, the patient should remain reasonably motionless during the x-ray exposure so that the radiologic technologist can obtain a sharply focused image. This is true for tomography, as well. To create the blurring effect, the x-ray tube and image receptor are set into motion. The tube/receptor motion takes place during the exposure while the patient lies completely still on the x-ray table.

Over the years, several types of tube/receptor motions have been utilized to obtain this blurring effect, including circular motions, vertical motions, and even spiral movements. Today, however, the preferred method is one in which the x-ray tube moves in a straight line, parallel to the tabletop and film receptor. This particular type of tomography, termed ***rectilinear tomography,*** or simply, ***linear tomography***, has proven most effective in producing quality tomograms. Furthermore, a linear tomographic unit is simplest in design and cheapest to manufacture.

The principles of tomography are more easily understood when visualizing the tube/receptor assembly as a fulcrum moving about a pivot point (see Figure 41). The tube and receptor (i.e., table bucky) are physically connected to each other and pivot about a central point located on ***a tomographic table***. The tomographic table is specifically designed to accommodate this modified tube/bucky assembly. In the most basic design, the x-ray tube is mechanically attached to the receptor with a solid metal ***connecting rod***. A point located at distance ***halfway*** between the film and x-ray tube becomes the pivot point, or ***fulcrum.*** When the x-ray tube is moved to the right (viewed from the front of the table) the bucky will move in the opposite direction and ***at the same speed*** as the x-ray tube.

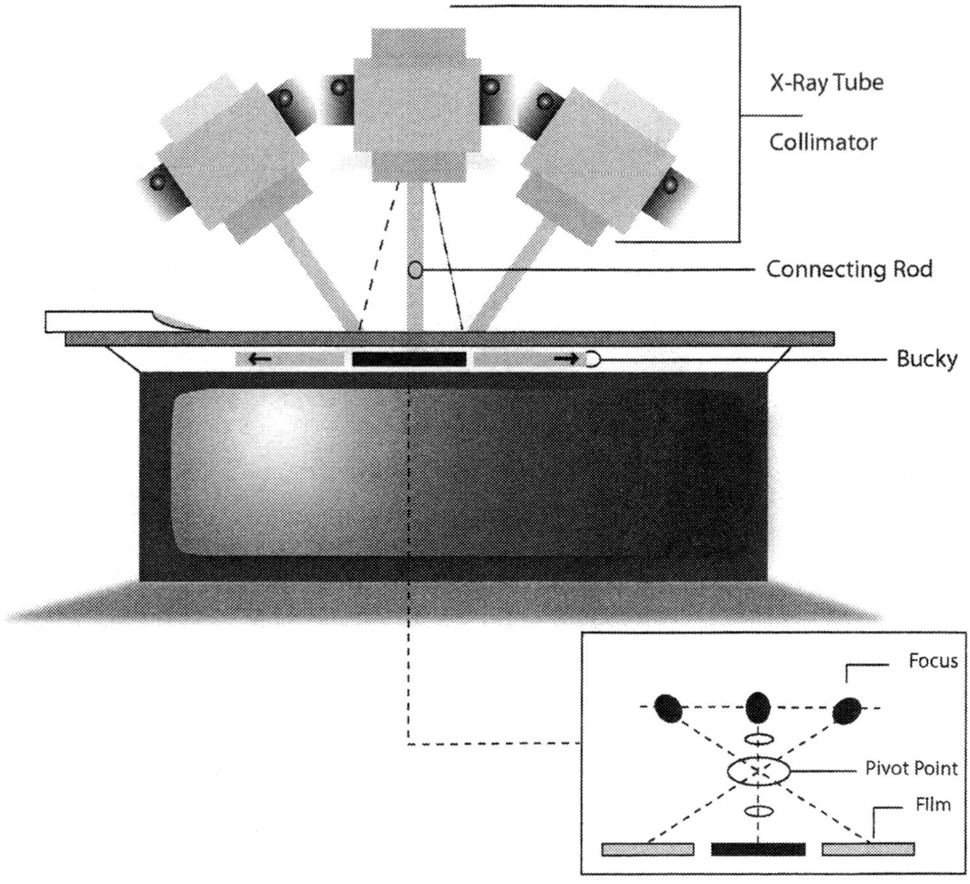

Figure 41. The tomographic table.

In tomography, *the x-ray tube and receptor must move simultaneously, at the same speed, and in opposite directions while the x-ray beam remains centered on image receptor.* Since radiation travels in straight lines through air, the connecting rod can be used to represent the central portion of the actual x-ray beam. From the diagram (Figure 41), it can be seen that the beam remains centered on the film throughout the entire range of motion.

The pivot point is the exact ***point of focus*** of the x-ray beam. Only objects located at the fulcrum will cast their shadow onto the center of the film during a tomographic exposure sweep. Structures that lie immediately above or below this point of focus will be blurred because their shadows will move at different speeds in relation to the speed of the shadow of the object located at the focus. From the inset of Figure 41, it is clear that a structure located at any point above the plane of the focus will cast a shadow that moves faster than the shadow of the structure located in the focus plane; the shadow of structures below the focus plane travels slower (a line drawn from the x-ray tube focus through the upper or lower structure will demonstrate this phenomenon). Both upper and lower structures will appear blurred on the film.

In a tomographic study, the pivot point is positioned at the plane within the body in which the desired structure lies. In this way, only that structure will appear focused on film even if surrounded by similar tissue. That plane, which is parallel to the film plane, is referred to as the ***objective plane.*** The pivot point (or focus) can be moved up or down through any objective plane within the patient's body. Specifically, if a patient is lying on the x-ray table in the supine position, the pivot point can be raised to the height of the sternum, or it can be lowered to the vertebral column located just a few millimeters above the table top.

When preparing for a tomogram, the radiographic technologist must consider two important variables: the exact ***location*** of the structure and the desired ***thickness*** of slice. First, a measurement is taken to determine the exact location of the desired section of the structure to be imaged. The technologist uses this measurement to set the ***tomographic height.*** The tomographic height setting is, in effect, the fulcrum height and pivot point. The technologist sets the tomographic height by moving the fulcrum to the measure distance. Now the pivot point is positioned precisely at the level of the desired anatomy and will not move during the exposure. With early design tomographic tables having a ***fixed fulcrum,*** the table height is moved up or down to adjust the tomographic height.

With tomographic height set, the technologist must adjust for how thick a section of anatomy is required for the study. The ***tomographic thickness,*** prescribed by the radiologist, will vary depending on the type of examination. Tomographic thickness is, in effect, the area of greatest sharpness within the imaged structure. When viewing a tomogram, objects very close to the tomographic height setting (pivot point) will appear relatively sharp. Moving away from the focus, image blurring gradually increases to a point where the anatomy becomes so blurred that it appears as background "noise." The tomographic thickness can be set so that the entire structure appears sharp on the film, or slice thickness can be set to view a section as thin as a few millimeters.

In reverse terms, increasing the amount of blurring of tissue located immediately above and below the desired structure will create the effect of a thinner section. Decreasing the amount of blurring will result in a thicker section that will remain in focus. *Slice thickness is controlled by the amount of blurring that is created.*

The way to control section thickness is by varying the distance that the x-ray tube travels during a tomographic excursion. Essentially, ***the greater the movement of the x-ray tube the more the blurring effect is exaggerated.*** When the x-ray tube is directly overhead, for example, the tube/bucky axis is perpendicular to the table, or at zero degrees. In this setup, which is normally used in general radiography, the focus approaches infinity. During a tomographic sweep, however, the tube/bucky assembly will begin at some point towards the head end of the table, for example, and move an equal distance through the zero point to a point at the foot end of the table. It is this motion that creates the blurring effect and by controlling the amount of motion, we can accurately control slice thickness.

In order to maintain exact focusing of the x-ray beam on a given structure during a tomographic sweep, the x-ray tube will have to rotate as it travels along its path. As the x-ray tube travels along the length of the table, it is rotated to remain aligned to the bucky. Moving the tube towards the head end of the table a given distance will cause the x-ray tube to rotate to a corresponding angle, for example, –20 degrees. Moving the x-ray tube in the opposite direction (towards the foot end of the table) an equal distance from the zero degree perpendicular will cause the x-ray tube to pivot to an angle of +20 degrees.

In tomography, instead of adjusting the sweep distance, the tomographic thickness is set by adjusting the tube angle, or ***tomographic angle.*** Consequently, at large tomographic angle settings, the x-ray tube will travel a significant distance, which will create an image of a thin section of anatomy. Smaller tomographic angle settings result in little movement and thicker sections (see Appendix G).

For example, starting at an angle of minus one degree, the tube must travel one degree from the head end of the table to return back to zero, then must travel through zero to plus one degree. The total movement is two degrees. Since this motion is minimal, most objects in the path of the beam will be sharply focused, similar in appearance to a general radiograph. Smaller angle sweep settings of ten degrees or less result in a relatively thick section or ***zone*** being imaged. A tomographic technique using small angles is termed ***zonography,*** which is very useful when the obscuring structures are far from the objective plane, or when the entire organ (kidney, for example) is to be imaged.

Moving the starting point to 20 degrees (head end), the x-ray tube now travels from minus 20 degrees through zero to plus 20 degrees for a total sweep of 40 degrees. The tube/bucky assembly now moves a significant distance, resulting in a greater amount of blurring. By increasing the tomographic angle setting even further to 40 degrees, the tube will travel a much greater distance along the table for a total sweep of 80 degrees, which will create an extreme amount of blurring. In this case, only a very thin plane measuring a few millimeters will be in focus on film.

Thus linear tomography allows for complete control of the slice depth and slice thickness. In tomography, only a few standard tomographic angles have proven effective for most studies. As a rule, a conventional tomographic unit will provide settings for 10 degree, 20 degree, and 40 degree tomographic angles.

Another parameter in tomography that must also be considered is the speed at which the tube travels during a tomographic sweep. The ***tomographic speed*** is set by the technologist to accommodate the individual patient application. Every tomographic unit provides a ***fast*** and a ***slow*** speed setting for x-ray tube/receptor movement. Fast tomography techniques are most often used when short exposure times are needed, such as for pediatric cases, or when the patient is experiencing pain and cannot remain motionless for long periods of time. The faster speed helps reduce patient motion which can detract from film quality.

Slow tomography sweep speeds are used under several conditions. First of all, many units are installed in rooms with limited space. Because of the momentum developed during a tomographic sweep, a fast excursion requires a greater braking distance simply because it takes longer for the tube to come to a complete stop. Therefore, in tomographic rooms with limited space, only the slow speed can be used. In addition, there is less vibration and smoother motion at slow speeds, which helps produce a better quality film. Ideally, slow speed should be used whenever possible.

There are, however, a few issues concerning the images produced by linear tomographic techniques that should be mentioned at this time. In general radiography, the x-ray tube is positioned directly over the organ to be imaged in an AP/PA projection, and the tube remains motionless during the exposure. This setup will always produce the sharpest image on film. Since the tube/bucky assembly is in constant motion during the tomographic exposure, the resulting images will be slightly distorted and minor streaking may appear on the film. Furthermore, in some instances, the final image can appear elongated in the direction of tube motion. These detracting effects to the image are a direct result of the tomographic process and are, unfortunately, unavoidable. Also, the images of other structures within the body can appear superimposed on the desired structure, depending on how close they are in respect to that structure. Needless to say, it takes an experienced eye to properly read a tomogram.

TOMOGRAPHIC EQUIPMENT

A tomographic unit is, in essence, a slightly modified general radiographic unit. The equipment in a tomographic suite is identical to that found in a general radiographic room with important exceptions found in three areas: ***the x-ray table, the tube support,*** and ***the x-ray generator.***

The Tomographic Table

As discussed in Chapter IV, the radiographic table is a specialized table that aids in patient imaging. In addition to the standard mechanical devices found in all radiographic tables, a tomographic table must include a mechanical means of connecting the x-ray tube to the image receptor. This physical connection ensures that the x-ray tube and image receptor move simultaneously and at the same speed. Additionally, tomographic tables also require an electromechanical means to control the precise movements of the tube/receptor assembly. A motor drive assembly is required, as well as the associated electronic circuitry to control the tube/receptor motion.

In most instances, the x-ray tube and image receptor (commonly a table bucky) are physically connected by either a rigid column, or by a simple ***connecting rod.*** The use of a connecting rod is often preferred since it is removable, allowing the equipment to be used for general studies. Many versions of connecting rods are currently in use. However, in the most basic design, the tube and receptor are linked together with a steel rod or bar. In this design, the connecting rod is attached to the table and x-ray tube prior to the tomogram and then can be removed and stored within the x-ray room for general radiographic studies. Some equipment designs allow for the rod to be permanently attached to the x-ray tube assembly. In this case, the rod is disengaged for general use.

The connecting rod attaches the table bucky to the tube support, a ***tubestand*** or an overhead ***tubecrane,*** and is commonly held in place with either a spring pin, or locking knob. The tube support must have a means of sensing the attachment to the table, and the additional circuitry that disengages the x-ray tube's lateral and rotational electromagnetic locks when connected. Once the connecting rod is secured, the tube/bucky assembly move as one unit. A motor drive system, located within the tomographic table, now controls the tube/bucky lateral motion. The motor drive assembly, which includes a motor and its associated gears and linkages, allows for precise positioning and braking of the tube/bucky assembly.

A ***control panel,*** usually attached to the table, is energized to engage the gears and controls all tube/bucky movements. Specifically, the control panel provides the tomographic technique selections for tomographic angle, sweep speed, and slice thickness. In addition, all tomographic controls provide a test switch that allows the operator to perform a tomographic "test" sweep without the use of x-rays.

The tomographic table must also have a means of identifying the position of the x-ray tube and the bucky. The easiest (and cheapest) way to detect and control this motion is by means of a series of strategically placed ***microswitches, magnetic switches,*** or ***photo sensors.*** Located just below the table top, these sensors detect the tube/bucky position and can be used to control the length of the tomographic excursion. Furthermore, elevating tables require an additional ***exposure interlock*** to indicate that the table is at the correct exposure height.

As will be discussed later in the chapter, the mechanical linkages between the x-ray tube and image receptor will wear over time and can become a source of imaging problems. In response to this "weak link" in the system, newer equipment designs have eliminated the connecting rod altogether, using instead a ***fulcrumless system.*** Fulcrumless systems utilize separate motors for x-ray tube and bucky movements, eliminating the need for any physical connections between the components. The individual motors are controlled by software, which provides accurate and smooth operation.

The X-ray Generator

The x-ray generator used for tomography is the same as that used in general radiography, with a few additional specialized components. In fact, many generators used for general radiography can be easily upgraded for special radiographic applications such as for tomography. The open design is a cost effective measure used by manufacturers. Generally, a tomographic x-ray generator will include a special function switch on the console, which enables the tomographic mode of operation. When in ***"Tomo" mode,*** the appropriate x-ray tube and receptor are activated. Tomographic generators also provide a wider range of mA selections.

Another consideration for tomographic generators has to do with the extended exposure times needed for an exam. During a tomographic excursion, the x-ray tube can move a distance of several feet or more as it travels from the head end to the foot end of the radiographic table. More importantly, x-ray radiation is being emitted during most of that excursion: exposures can be as long as five seconds in duration! Therefore, the x-ray tube and generator must be able to accommodate these longer exposure times. One way to ensure that the heat and exposure limits of the x-ray tube will not be exceeded is by the use of ***very low mA technique settings.*** Consequently, all tomographic x-ray generators must have the capability of providing these significantly lower mA settings.

The x-ray generator should provide a selection of 25 mA and below on the small focus of the x-ray tube. At these lower mA settings the longer exposure times can easily be achieved without the risk of damaging the x-ray tube. Accordingly, tomographic techniques commonly provide settings of 5 mA, 10 mA, 15 mA, and 25 mA.

OPERATION

For a tomogram, the patient is positioned on the table top so that the desired anatomy is centered directly over the image receptor. The plane of the structure is measured (in reference to the table top) and the fulcrum is set to that distance. A ***scout film*** (i.e., normal AP/PA radiograph) is taken to ensure that the kVp technique setting is correct, and also to verify positioning. The operator then selects "Tomo" mode on the x-ray console and also programs the tomographic angle and speed on the tomographic control. At this point, a ***test run*** (i.e., tube excursion without x-ray radiation) can be performed to insure that the unit is working as programmed and that there are no obstructions in the path of the x-ray tube/bucky assembly.

Next, the x-ray tube is moved to the starting point for the tomographic sweep. The exposure is initiated at the x-ray control via the handswitch which sets the tube in motion. Since ***smooth movement*** is essential to obtaining a quality tomogram, the x-ray tube should be moving at constant speed prior to emitting radiation. Once the tube reaches the programmed tomographic angle, the exposure will begin. The exposure terminates as programmed by the x-ray timer, or when the tube finishes its sweep and braking is initiated.

CALIBRATION

To calibrate a tomographic unit, the service engineer should first complete all of the calibration procedures required for general radiographic equipment as described earlier in this book. In addition, the engineer should focus specifically on two areas: the lower range of mA calibrations and the tomographic table adjustments.

The service engineer should take extra care when calibrating the lower mA stations to ensure the tube current waveforms at 5, 10, 15, and 25 mA are stable for the entire duration of the exposure and that they remain as flat as possible. The reason for this is twofold. First of all, it is generally difficult to obtain a stable waveform at such low current settings because of the inherent characteristics of the x-ray tube. Second, with the longer exposures required for tomography (e.g., 1–5 seconds) there is a greater chance of waveform fluctuations as a result of fluctuations and noise occurring on the incoming line. The use of an oscilloscope when calibrating the mA stations is highly recommended to ensure proper adjustments. Once all of the mA stations are calibrated, the engineer should observe the mA waveforms on the oscilloscope during longer exposures to verify calibration.

The engineer should also concentrate on the calibrations of the tomographic table. For precise imaging, the following parameters should be calibrated: ***fulcrum height, sweep speed,*** and ***tomographic angle.*** Since the designs of tomographic tables can vary, general calibration principles are describes below.

The fulcrum height is adjusted so that when, for example, the fulcrum height indicator displays five centimeters, the true focus is exactly 5 cm above the table top. Essentially, what is being calibrated is the height indicator (i.e., analog or digital readout). The service engineer will measure above the table top a given distance and calibrate the readout to match the actual setting. The adjustment should be verified with a test film. To do this, a phantom should be placed on the table top and then exposed. Most ***tomographic phantoms*** are designed with incremental steps that are clearly marked with lead numeric markers. If a 5 cm fulcrum height is set, for example, then the number "5" will appear in focus on the exposed film and numbers "4" and "6" should appear blurred.

Sweep speed is adjusted in various ways. In basic designs, the speed is set by first measuring the actual time it takes to make a full sweep, and then by calibrating the sweep time to manufacturer's specifications. A stop watch is normally used for this calibration. Tomographic angles are measured, and then adjusted by changing the locations of the corresponding microswitches or electronic sensors. In many instances these angles are fixed and field adjustments are not required.

FAILURES WITH TOMOGRAPHIC UNITS

Generally speaking, whenever excessive movement is involved with radiographic equipment, predictable failures will occur as a result of those movements. If the x-ray tube is repositioned frequently, as it is in general radiography, the interconnecting cables that attach to the x-ray tube tend to be stressed. Because of the repeated motions associated with tomography, all of the interconnecting cables attaching to the tube support will receive similar stress. Eventually, these cables will become damaged or the connectors can become dislodged or broken. Accordingly, on every preventive maintenance, the service engineer should carefully inspect the condition of these cable harnesses, and the cables that attach to image receptor, looking for loose connectors and worn insulation.

Another common mechanical failure occurring with tomographic units is likewise a result of the routine movements of the tube/bucky assembly. Without question, the single most important criterion for tomography is that tube and receptor ***remain vibration free*** during the exposure. Specifically, the amount of "play" in the mechanical linkages must be minimal in order to produce a quality image on film. Play in the linkage causes uneven or ***lost motion*** that will result in unwanted blurring on the film. A loss of image quality is most noticed when the play develops at the fulcrum of the tomographic table, since the effects of lost motion will be magnified here. This undesirable blurring will also occur when the play occurs at the bucky linkage. Finally, lost motion has minimal effect when it occurs at the x-ray tube.

Because of the effects of lost motion, the number of linkages should be kept to a minimum with equipment design. Regardless, image quality problems often result from a worn or damaged linkage in the tube/bucky assembly. The engineer should, therefore, routinely inspect the tube/receptor assembly during a preventive maintenance inspection and should make the appropriate adjustments or repairs when necessary.

The microswitches used to detect the tube/receptor movements are also prone to failure especially with equipment that utilizes mechanical switches and detents. These components simply wear down with normal use and require frequent adjustments. Systems that employ magnetic or IR sensors are less problematic.

As mentioned earlier in the chapter, the best equipment designs utilize ***fulcrumless systems*** that employ separate microprocessor controlled drive systems for both the tube and the bucky movements. With these systems, the x-ray tube and image receptor do not have a physical connection. Therefore, there are no linkage-related problems. Fulcrumless systems have greatly reduced the potential for lost motion.

Chapter XII

MAMMOGRAPHY

OVERVIEW

Mammography is a specialized field of radiography developed specifically to help in the diagnosis of breast-related cancers. Other than skin cancer, breast cancer is the most common form of cancer among women and, as of this writing, is the second leading cause of death for women. Mammography x-ray systems are uniquely designed to produce high quality breast images that can be used for the screening and diagnosis of breast cancer.

Since first developed, mammographic equipment design has undergone significant changes–many improvements a result of the MQSA (Mammography Quality Standards Act), which, in 1994, established specific guidelines for mammography. Today, mammography is the first choice for cancer screening, having a sensitivity of approximately 90 percent (i.e., 10% of carcinomas are not detected).

Because of the composition of breast tissue, special techniques are needed to obtain diagnostic images. The quality of the x-ray beam, the type of film, the method of film processing, and even the reading room used by radiologists for diagnosing breast images must must be altered in order to visualize fine details of breast tissue. Moreover, stringent quality control measure are routinely taken to verify equipment calibrations, and to ensure consistency in film processing and viewing conditions.

Because of the many regulations involved with mammography, the service engineer must remain current on all MQSA guidelines so that he or she can service mammography equipment effectively. In addition, he or she should be completely familiar with equipment designs, noting how each component affects the final image. Finally, a general knowledge of breast anatomy is extremely beneficial for engineers, especially when discussing imaging problems (and solutions) with the hospital staff.

This chapter begins with an informative overview of breast anatomy and will describe specific techniques commonly used during a mammographic exam. Next, mammographic equipment design is discussed in detail, focusing on the components that help produce diagnostic quality images. An overview of calibration procedures is followed by a discussion of common failures that occur with mammography equipment. A final note on digital mammography completes the chapter.

THE BREAST

The human breast is comprised of a variety of soft tissue which presents some unique imaging challenges. The main structure is the ***mammary gland*** which consists of 15 to 20 lobes, each with numerous related ducts and lobules. The individual lobes of the mammary gland open into the lactiferous ducts, all of which converge at the nipple. The body of the mammary gland is imbedded within ***fatty tissue,*** is supplied by a network of ***blood and lymph vessels,*** and is supported by ***connective tissue.***

Furthermore, the range of tissue densities that can exist is even greater with malignant breast tissue. Both the ***radio-dense*** tissue (e.g., fibrotic tissue, glandular tissue, and malignant tissue) and ***radio-lucent*** tissue (e.g., fat or loose connective tissue) must be clearly visible on the mammogram so that the radiologist can make accurate diagnoses. Accordingly, a mammography system must provide very low kVp techniques, in the range of 20–40 kVp, at relatively high mA/mAs outputs. In addition, extremely small focal spots (0.1–0.3 mm) must be used to image the very fine microcalcifications.

MAMMOGRAPHIC VIEWS

Since the service engineer cannot directly observe a patient during a mammographic examination, it is beneficial to understand exactly what is taking place during that procedure. This knowledge is essential for troubleshooting mammography problems and also important when discussing image quality issues with the staff. When investigating imaging problems, the engineer must rely on the limited information provided by the technologist and from the clues discerned by reviewing a questionable film. Possessing knowledge of examination protocol gives the engineer a distinct advantage and also communicates service professionalism.

As a rule, a routine mammographic screening examination requires at least four films, two exposures per breast. All exams should include filming in two planes: the ***Mediolateral Oblique (MLO) view*** and the ***Cranial Caudal view (CC).*** These two ***standard views*** permit reliable diagnoses of breast abnormalities. ***Additional views*** should be used only when standard views are inconclusive. Some additional views include a ***spot compression view,*** a ***magnification view*** (see below), a ***90 degree lateral view,*** and a ***rolled view.***

The ***MLO*** projection is performed at 30 to 70 degrees inclination from the vertical and provides good visualization of the entire breast. Most carcinomas can be identified in this view. The MLO view is supplemented with the ***CC*** view, which allows full visualization of the entire body of the mammary gland. In a CC view, the x-ray tube is positioned directly overhead, with the beam traveling from superior to inferior.

Tissue compression is used with all procedures for reasons described later in this chapter. With standard views, the compression paddle must match exactly the size of the film being used and full field collimation is utilized so as to image the entire breast. However, occasionally there is a need to spread a small area of interest within the breast tissue in order to help differentiate superimposed structures. In this case, ***spot compression*** is used. The compression paddle (or cone) used in spot compression is much smaller than a standard compression paddle, measuring as small as 7 cm. In addition, spot compression is performed with the x-ray beam collimated down the area of interest only.

In addition to the spot compression view, a magnification technique can be used to help visualize a questionable area of the breast. When performing a ***magnification view,*** a magnification table is placed above the image receptor, raising the patient's breast to a fixed distance above the film plane. The gap created between the breast and film results in a magnified image of a specific area of the breast. The ***small focus*** of the x-ray tube is used with all magnification techniques and the exposure field is collimated to the specific area of interest. With magnification, the grid is always removed. Scatter radiation is reduced, instead, by the use of (1) an ***air gap,*** which is created within the magnification table and (2) by ***collimating*** "down" to the area of interest only. Spot compression is often used with magnification technique.

THE MAMMOGRAPHY UNIT

A mammography unit should meet certain requirements to produce a good quality ***mammogram.*** The unit must have sufficient power to image breasts varying in size and density, while using a minimum amount of radiation. It should produce sharp images of the extremely fine ***microcalcifications*** located within the breast tissue and must be also provide precise positioning at the specific angular projections (discussed above) known to facilitate good imaging of the entire breast tissue.

As mentioned earlier, ***tissue compression*** must be employed during each mammographic examination and all units should also provide a means of image ***magnification.*** Finally, mammography units should be made of quality components, capable of handling high patient loads yet be compact in design to facilitate installation in imaging suites with limited space.

A mammography unit is essentially a completely integrated x-ray system designed as a "stand-alone" unit. Each device normally found in a general x-ray room has been slightly modified and consolidated into one single unit (see Figure 42). The x-ray generator, x-ray tube, tube support, collimator assembly, filter assembly, compression assembly, and the image receptor are all part of the mammographic unit. In this chapter, the mammographic unit is broken down into three sections: the ***gantry assembly,*** the ***c-arm assembly,*** and the ***x-ray control console.***

The Gantry

The ***gantry*** is an integral part of a mammographic x-ray unit since it houses many of the system's components. It also provides the structural support for the ***c-arm assembly*** that contains the complete imaging chain. The gantry must be made of a rigid steel framework so as to sup-

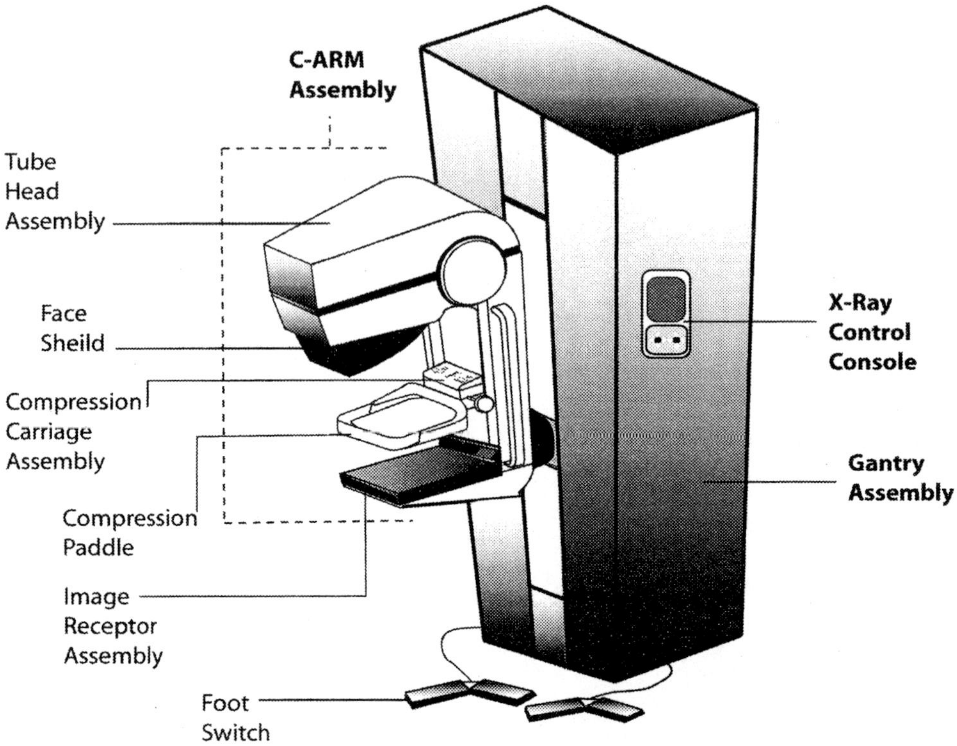

Figure 42. The mammographic system.

port the significant weight of the c-arm. Also, a solid structure is needed to accommodate smooth, precise movements of the x-ray tube and image receptor during the examination.

Included within the gantry are the motors, gears, and associated electronic circuitry needed to control the c-arm movements. All of the power supplies for the system, as well as the microprocessor and display circuitry are housed within the gantry. Finally, the gantry contains the complete ***x-ray generator.***

The X-ray Generator

For many reasons, the x-ray generator of choice used for mammography is the ***high frequency generator*** (see Chapter III). High frequency generators are extremely reliable, require input voltages commonly found in most hospitals and clinics, and, more importantly, take up very little space. Also, since mammography utilizes x-ray techniques that are comparably much lower than those used for general radiography (20–40 kVp @ 30–100 mA), the high voltage transformer is relatively small is size. This small footprint helps contribute to the compact design of mammography units.

The ***high voltage transformer*** is conveniently located in the base of the gantry, with the generator's primary components located on a chassis within the mainframe, or fastened to a side panel of the gantry. With this layout, the interconnecting cables are run internally, which allows for shorter cable runs and helps provide cleaner appearance of the unit.

The high voltage transformer is connected to the x-ray tube, often with a single ***high voltage cable.*** With the single cable design, only one insulated cable is needed to carry both anode and cathode voltages to the x-ray tube, which also aids in the compact design of these units. The standard "federal" connector, used in radiography, is, however, still used by some equipment manufacturers.

The C-Arm Assembly

With all mammography units, the x-ray tube is physically attached to the image receptor by a single rigid column utilizing a common ***c-arm design.*** Referring to the diagram (Figure 42), it can be seen that the x-ray tube is connected to the c-arm column at one end, and the image receptor at the opposite end, always at a ***fixed SID*** (typically 65 cm). The c-arm, supported by the gantry, can be raised or lowered to adapt to the individual patient height.

In addition, the c-arm can be rotated to the precise projection angles required to visualize the components of the breast. In effect, the c-arm can be rotated to nearly 360 degrees of angulation. The c-arm assembly contains the entire imaging chain which includes the ***x-ray tube, filter assembly, collimator assembly, compression paddle assembly,*** and ***image receptor.***

The X-ray Tube

The design of a mammographic x-ray tube is unique in a number of ways. These special x-ray tubes are physically smaller, overall, than those used in general radiography and are designed to produce the low energy x-ray spectrum needed to create a high contrast image of breast tissue. Mammography tubes are designed to operate at voltage potentials in the range of 20 kVp to 40 kVp. However, relatively high mA's, typically @ 30–100 mA, are required to image dense or malignant breasts. Thus mammography tubes must have high output ratings (i.e., 8 mr/mAs or greater).

Two designs are commonly used by x-ray tube manufacturers: ***grounded cathode*** and ***grounded anode*** x-ray tubes. The grounded cathode tube is commonly found on equipment because it is simpler to manufacture and therefore cheaper to produce. However, a grounded anode tube is the best design, overall, mainly because it greatly reduces off-focus radiation which can effect image quality. The grounded anode tube is more difficult to manufacture, though, and can cost three times as much as a cathode grounded tube. In either case, the cost of manufacturing mammography x-ray tubes is much higher than for tubes used in general radiography, starting at $10,000 as of this writing.

Another consideration with x-ray tubes relates to equipment usage. In a typical mammographic diagnostic study, ten or more exposures may be required, depending on the condition of the particular patient. The x-ray tube must be able to dissipate the excessive heat that is generated during a mammography examination. Consequently, most mammography x-ray tubes are designed with a relatively large housing in relation to its insert, which is completely surrounded by insulating oil. The large tube housing, used in conjunction with a vented hood surrounding the x-ray tube, and a relatively high volume blower (i.e., box fan) all help to dissipate the heat generated during the examination.

As stated earlier, a single high voltage cable is frequently used to carry the kilovoltage to the x-ray tube. The ***ALDEN*** cable, originally designed for use in the power supplies for CRT-type display monitors, plugs directly into the anode end of the x-ray tube housing. With this single cable design, a second cable is needed to supply power to the filament and tube safety circuits. This second, low voltage cable runs from the high voltage transformer to a separate connector found on the x-ray tube housing. By having the two power connectors located at the anode end of the housing, the tube ***glass insert*** can be positioned as close as possible to the cathode end of the housing, which places the exit port closest to the patient. The port should be as close as possible to the patient's chest wall so that the entire breast can be imaged.

Another important feature of mammographic x-ray tubes is the use of a very small ***focal spot size.*** The extremely small focuses are required to image the microcalcifications and fibrotic strands in the breast. Common focuses used in mammography range from 0.1 mm to 0.3 mm (maximum, both dimensions). These focuses are very small in comparison to those used in general radiography, which usually range from 0.6 mm to 1.5 mm.

The ***target*** is also specially designed to produce an x-ray beam consisting mostly of low energy x-ray radiation. Though some x-ray tube manufacturers use a double surfaced anode (i.e., molybdenum/rhodium) in their design, most targets are primarily made of molybdenum, which produces a higher portion of low energy x-ray radiation (17.5–19.6 KeV). Molybdenum targets, when used with specifically designed filters, allow for effective control over the radiation spectrum required for breast imaging. Molybdenum is optimum for mammography, however it has a lower melting point which will cause the target to erode over time.

As with all x-ray tubes, the electrons colliding with the target produce a ***heel effect.*** Since the thickness of the breast is greatest at the chest wall, the target should always be oriented to the nipple end of the patient.

The window or ***port*** of a mammographic x-ray tube is also made of a special material that allows soft x-ray radiation to pass through to the patient. The lower energy radiation will be differentially absorbed within the soft tissue of the breast providing a high contrast image. Beryllium windows are commonly used in these x-ray tubes.

The Filter Assembly

The modified (soft) x-ray beam that exits the port of the mammography tube must pass through additional ***filtration*** to create an x-ray beam that will produce a diagnostic quality image of the breast. With mammography, one filter is not sufficient to image all types of breast tissue, which can vary significantly due to the age and condition of the patient. For example, the breast of a normal adolescent female will generally contain much less fatty tissue than that of mature female. Even more, a breast with abundant fibrotic tissue or one with an implant will require different levels of radiation to obtain a satisfactory image.

Because of the wide range of breast tissue, there is a need to alter the beam quality precisely for the particular breast being imaged. Also, since a much "softer" x-ray beam is needed for mammography, the typical minimum HVL is much lower than for general radiography (i.e., 0.3 mm to 0.5 mm of aluminum vs. 3.0 mm to 5.0 mm of AL for radiography). In mammography, the quality of the beam is modified by varying the kVp applied to the x-ray tube and, by ***adjusting target/filter combination.***

Rather than an aluminum filter, which is used for general radiography, most mammography units utilize two standard filters to adjust the beam quality more accurately. For mammographic applications, both ***molybdenum*** and ***rhodium*** filters are used. Molybdenum filters are used for small to normal density breasts, whereas rhodium filters are used for large or dense breast, especially when higher kVp's are needed. Both filters will suppress the very low-energy components of the x-ray spectrum that do not contribute to the image while reducing higher energy radiation above the ***K absorption edge*** characteristic of each filter. Each filter type has a specific limit particular to that filter. Above this limit, radiation will be absorbed. That limit is termed the k absorption edge. In effect, by the use of selective filtering, only a narrow (useful) band of radiation will reach the patient.

The filter, as in general radiography, is always positioned as close as possible to the x-ray tube port. However, since ***two different filters*** are used, a mechanical means of changing these filters is needed. Most mammography units utilize a ***motorized filter assembly*** that allows the operator to change filters from the control console. The latest equipment designs provide automatic filter selection based on the technique used. An "auto-filter" mode of operation is preferable since it greatly reduces any chance of operator error.

The Collimator

The collimator found on most mammography units is of two common types. Earlier vintage machines utilized a ***diaphragm assembly.*** In this design, a diaphragm is precisely matched to the specific bucky assembly being used. The diaphragm is inserted into a slot located near the x-ray tube port in preparation for the exposure. By using only that diaphragm, the operator is assured that *the light and x-ray field will exactly match the film size.*

Newer units have a ***motorized collimator assembly.*** With motorized collimation, each of the four blades of the collimator has a separate drive motor that can be individually adjusted. During calibration, these blades are set to match the size of the film and bucky being used.

Motorized collimators are not only convenient but also provide more precise collimation of the radiation field.

The ***collimator lamp*** used in mammography is similar to that used in general radiography. The lamp should illuminate the entire exposure field to within two percent of SID and must meet the minimum light output (160 lux minimum/15 foot-candles) required by MQSA standards. Mammography x-ray units must also have a precise means of adjusting the size and position of the light field. When the lamp has burned out, it should be replaced by a qualified service engineer, only, because of the series of adjustments, verifications, and documentation that are required.

A ***face shield*** is an important component of the mammographic unit because of the proximity of the patient to the x-ray tube. Usually attached to collimator assembly, the face shield is made of a hardened, lead-lined acrylic. MQSA requires that the face shield be in place during the examination because it greatly helps in reducing the amount of scatter radiation directed towards the patient.

The Compression Assembly

To produce a diagnostic quality image of the breast, a method of ***tissue compression*** must be used. Breast compression accomplishes several important goals, all of which help improve the image quality of a mammogram. Compression reduces the overall thickness of the breast so that the beam will pass through less tissue, thereby ***reducing scatter radiation.*** From the chapter on imaging we know that by reducing the amount of scatter radiation, there is less radiation dosage absorbed by the patient and, also an improvement in image contrast results. The use of compression also ***reduces the OID*** (object to image distance), greatly improving image resolution.

There are numerous other benefits to breast compression as well. When breast compression is used, the healthy tissue will spread more easily than will lesions or masses. As a result, *breast compression helps visualize true masses and reduces the likelihood of false identification of lesion.* Finally, compression helps ***reduce patient motion*** and ***suppresses arterial pulsations*** which helps improve image sharpness.

Because of the many advantages of utilizing compression, all mammography units provide both manual and motorized tissue compression (Figure 42). Motorized compression is initiated either by a footswitch located at the base of the gantry or, by a switch located on the c-arm. In preparation for an exposure, the technologist will position the patient's breast directly onto the bucky surface and then lower a compression paddle down onto the breast to a ***precompression force*** level. Next, the technologist will manually compress the breast by turning the compression knob until the desired amount of compression is achieved.

The compression assembly consists of a ***compression drive,*** a ***compression carriage,*** and a ***compression paddle.*** A compression motor drives the compression carriage (with paddle) up and down the c-arm via a chain or belt. Once in position the ***motor brake*** prevents any backlash in the compression carriage. Also, for patient safety, the compression drive includes a ***clutch assembly,*** designed to prevent overcompression during manual compression.

The motorized precompression force is adjustable in the range from fifteen to thirty pounds of force. For manual compression, the compression clutch will disable the compression drive when a maximum compression of typically sixty-five pounds is reached. Most mammography units provide a means of indicating the ***compression height*** and ***compression force*** used during the exposure, since these values are crucial to image analysis. The latest designs incorporate an ***automatic compression release.*** Here, once the exposure has been completed, the compression paddle retracts automatically, thus aiding patient comfort.

The ***compression paddle*** should meet certain guidelines. Each paddle must match the exact size of the bucky that is being used. Since the paddles are interchanged frequently throughout the day, they should be designed for quick and effortless removal and replacement. Most importantly, the paddles should be strong enough to withstand a significant force, yet must not alter the x-ray beam in any way. Consequently, most paddles are made of acrylic. With proper care, these compression paddles could last a number of years. However, repeated overcompression applied to the phantom (during QC testing) will cause a premature failure, usually indicated by cracks forming along the chest wall ridge.

The Image Receptor

The image receptor assembly used in mammography is similar in function to the one used for general radiography but has been modified for breast imaging. The image receptor includes the ***x-ray grid system,*** the ***film cassette holder,*** and the ***automatic exposure control (AEC) circuit.*** In addition, a ***magnification table*** is an integral part of all mammography units.

As stated earlier in this chapter, the x-ray tube and image receptor are physically attached to the c-arm at a fixed SID. The receptor, or bucky, is attached to the bottom or foot end of the c-arm, providing a solid surface on which to position the breast. The bucky surface must be rigid and free of defects that could cause artifact on the image. Two common bucky sizes are used in mammography: the 18 cm × 24 cm (8″ × 10″) and the 24 cm × 30 cm (10″ × 12″) buckys. The 18 cm × 24 cm bucky, commonly called the ***small bucky,*** is used first, if possible. The small bucky uses a smaller field size, which reduces the amount of scatter radiation. The ***large bucky*** is used for breasts that cannot be fully imaged on the small bucky.

Similar to the receptor used in general radiography equipment, the bucky houses the x-ray grid and grid drive mechanism. As with all buckys, the x-ray grid moves continuously during the exposure so that the image on film is free of grid lines. For mammography, two special grids are utilized to improve image quality: ***linear grids*** and ***cellular grids.*** Linear grids are most commonly used by equipment manufacturers. Because of the absorption characteristics of breast tissue and the increase in radiation dose associated with higher grid ratios, linear grids should always have a low grid ratio. Consequently, grid ratios of 4:1 or 5:1 (lines per inch) are commonly used in mammography units. Cellular grids, which are also highly effective at reducing scatter radiation at very low kVps are currently being used by only two equipment manufacturers.

The bucky assembly also includes the film cassette holder. A film cassette loaded with fresh (i.e., unexposed) film is positioned into a slot located in the bucky in preparation of exposure. Usually, a mechanical latch centers the film cassette within the radiation field and locks it into place. Once the exposure is taken, the cassette is removed by manually pressing a latch release.

Finally, the bucky table contains the AEC detector. The detector is an ***exit-sensing device*** which senses radiation that has passed through the breast, grid, and film cassette. Once a calibrated amount of radiation reaches the detector it will generate the exposure stop signal used to terminate the exposure. The detector is movable, so that the technologist can position it directly under that portion of the breast that will achieve the best image density. Positioning of this photocell is vital to obtaining good images–incorrect positioning is often a cause of image density problems.

The X-ray Control Console

The x-ray control console of a mammography unit is most often located in a corner of the x-ray room but can also be attached directly to the gantry. In either case, the control is secured behind a lead shield which protects the operator from radiation exposure. The x-ray control is similar in design to all other x-ray control units, allowing the operator to select the x-ray exposure technique (e.g., kVp, mA, and time), the mode of operation (e.g., manual vs. automatic), the filter type (e.g., molybdenum or rhodium), and to initiate the exposure sequence.

In addition, all of the patient data is entered via the control console ***keyboard*** and can be subsequently downloaded to the console hard drive, or the data can be stored on a floppy disk. The information is also recorded on the film label, so that each film contains the name of the patient, the exposure technique, and additional information such as the compression height and force. All films are labeled to accommodate accurate record keeping, another requirement of MQSA.

X-RAY FILM, FILM CASSETTES, AND FILM PROCESSING

The x-ray film used in mammography differs in a number of ways from the film used in general radiography. The main difference between the two types of film lies with the film emulsion. Whereas in general radiography a double-emulsion film (i.e., emulsion on both sides of the film) is used, mammography requires the use of ***single-emulsion film.*** Double-emulsion film (used with dual screens) is very effective for general radiography but does not produce an image sharp enough to visualize the fine microcalcifications within the breast. Consequently, all mammography designs use single-emulsion film.

Also, single emulsion film must be used with film cassette containing a ***single intensifying screen.*** To produce the sharpest image, the emulsion side of the single emulsion film must be in ***direct contact*** with the single ***intensifying screen.*** In addition, only a ***back screen*** configuration is used. With a back screen configuration radiation passes through the film to the screen located behind the film. This configuration ensures that the greatest intensity of light exiting the screen surface travels the shortest distance possible to expose the film, thus achieving the high resolution image.

As with general radiography, the correct film/screen combination is required to obtain the best quality images. The film used in mammography must produce the highest contrast possible over the density ranges that produce best visualization of tissue. The optimal density range for mammography is 0.6 D to 2.2 D (2.8 D with bright spot light).

Once a set of films has been exposed during the mammographic examination, the technologist will send them to the darkroom for processing. Again, the requirements for film processing are quite different than those used for general radiography. As a rule, a film processor ***dedicated*** to mammography should be used. In addition, processors must be actively running throughout the day to ensure stability. Usually a recommended minimum number of films (e.g., 20 films per day) must be run through the processor to ensure that fresh chemistry is being circulated throughout the processor.

Since any number of parameters can be affected by film processor performance, strict quality control standards are employed here as well. It is recommended that the imaging center follow the film manufacturer's guidelines for processor quality control, which requires the technologist to record phantom readings routinely, as well as running sensometric strips.

VIEWING CONDITIONS

The environmental conditions of the film reading room, important for general radiography, are crucial in mammography. Ideal viewing conditions are absolutely essential for the proper diagnoses of mammograms. Consequently, strict and specific guidelines have been established to ensure that optimum viewing conditions exist at every mammography center. The ultimate goal is to have identical environments in all reading rooms so that a film viewed at one facility will appear the same when "read" at different imaging center.

As in general radiography, the light output of a veiwbox must be of a specific color and intensity. However, the light output needed for viewing is much higher (2000–3000 cd/m^2 vs. 1400 cd/m^2 for general radiography). Viewboxes must be cleaned and inspected on a routine basis. When reading films, the radiologist should ***mask*** the area immediately around the film to aid in an accurate diagnosis of the image. Furthermore, a ***bright spotlight*** must be available in the reading room. The high intensity output of the spotlight (20,000 cd/m2) aids in the visualization of the fine details in areas of higher exposure (2.8 D to 3.0 D maximum density). A magnifying lens of at least two times magnification, an MQSA requirement, should be used during film evaluation.

INSTALLATION OF MAMMOGRAPHY EQUIPMENT

The installation of a mammography unit is much simplified over that of general radiographic equipment. The stand-alone design eliminates the elaborate cable runs and most of the alignment procedures needed for radiography. For installation, the assembler will locate the unit within the mammographic examination room as specified by the room drawing and secure it to the floor. A dedicated and clean power line is connected directly to the power circuits of the gantry, and one or more interconnect cables are run between the gantry and control console.

Today, most units are calibrated at the factory for the specific x-ray tube installed in the unit. Usually a functional test is all that is required to verify proper operation and to ensure that the unit was not damaged in any way during shipment. However, the service engineer should always verify all calibrations, especially the mA calibrations and compression adjustments. Otherwise, the major part of the calibration procedure consists of the setup and calibration of the automatic exposure system (see below).

QUALITY ASSURANCE IN MAMMOGRAPHY

Before proceeding with calibrations it is important to understand the source of the calibration standards. In the United States, all aspects of mammography are prescribed by the Mammography Quality Standards Act (MQSA). MQSA sets guidelines for all aspects of mammography, including standards for radiation producing equipment, film, and film processing, viewing conditions of the reading room, and even for patient record keeping. These guidelines, enforced by the FDA (Food and Drug Administration), were created to help ensure that patients receive consistent quality care at all mammography centers throughout the country.

In order to operate legally, every mammography facility must be accredited by the American College of Radiology, or ACR. Designated by the FDA, the ACR establishes standards and practices that are to be strictly followed. Each mammography center must perform ***quality assurance (QA)*** procedures regularly on all components of the mammographic system. Specifically, the darkroom must be inspected daily for cleanliness, and film processor quality control is to be performed on a daily basis. Phantom images must be evaluated every week to confirm proper equipment operation. In addition, film cassettes must be cleaned and tested and viewing conditions are checked on a weekly basis.

Mammography units are checked weekly by the use of a special mammography ***phantom.*** The phantom used for mammography must be approved by the ACR. A typical ACR phantom, made of acrylic, contains a wax insert with fibers, specks and masses. The phantom must simulate a 4.2 cm compressed breast and the phantom image should reveal at least four fibers, three masses, and four specs. MQSA/ACR guidelines for phantoms can be be found at each facility and are available on the internet.

Equipment calibrations are also set by the MQSA. When new equipment is installed, the service engineer must follow the established guidelines for calibration accuracy. Once a unit is installed, it must be thoroughly tested by a radiation physicist who will ensure the all guidelines and standards are met.

PREVENTIVE MAINTENANCE AND CALIBRATION

Mammography units are calibrated to meet all of the manufacturer's specifications, which must also comply with MQSA guidelines. Mammography systems should be inspected and calibrated at least two times per year. When servicing equipment, the engineer can follow the general guidelines for preventive maintenance and calibration that are used for radiographic equipment. Specifically, the service engineer should clean and inspect all of the mechanical assemblies, verifying correct operation. Also, the x-ray generator can be calibrated using the principles described earlier in this book.

Additionally, the compression assembly must be calibrated to achieve safe and effective compression forces and to display the correct compression height. The collimator calibrations are similar to those found in general radiography but a bit more elaborate because of the critical tolerances required for mammography.

Finally, the automatic exposure circuits must be checked and calibrated. With mammography, this is usually the most crucial and most time consuming adjustment. During the AEC calibration, the engineer will take exposures using varying thicknesses of acrylic and then make adjustments so that the densities will track consistently over the complete range of thicknesses (i.e., 2 cm to 8 cm of acrylic). For mammography, the AEC circuit must track over the entire range of kVp's by no more that +/– 0.3 density (+/– 0.15 D for small bucky). The engineer should use great care during this calibration since it will ultimately impact the quality of the final images.

FAILURES WITH MAMMOGRAPHY UNITS

Mammography equipment will fail in ways similar to and with the same frequency as general radiographic equipment. The difference with mammography service lies in the ***servicing approach.*** The main purpose of a mammography examination is cancer screening. Patients who arrive for examination are often anxious, worried, intimidated, and, at the very least, uncomfortable. The service professional should be conscious of the sensitive nature of mammography service and should always use discretion when on service calls.

A good practice when arriving for service is to register at the reception area and then remain in a designated waiting area until permission is given to enter the mammography suite. The engineer's service tools and test equipment should remain outside of the imaging center as well, since their presence could be a distraction to the patients. As always, a professional appearance and common courtesy go a long way.

When servicing mammography equipment, the engineer can utilize the troubleshooting practices outlined earlier in this book. In fact, experience in general radiographic service is of great advantage when tackling the service issues specific to mammography. Many failures occur with mechanical assemblies such as electromagnetic locks, motor drives, control switches, and other common x-ray components. Failures in these areas can be corrected using general repair skills. Below we will concentrate on the failures that are unique to mammography equipment, detailing ways to isolate their causes.

It has been the experience of the author that most mechanical failures with mammography units commonly occur in two areas: the compression drive assembly and the image receptor. Because of the amount of force used during compression, the major components of the ***compression drive*** tend to fail over time. Specifically, drive chains will break, gears become worn, and the clutch and drive motor will eventually fail. The compression paddle, which is removed and then reinserted for each case, will require repair or replacement after a period of time. Also, if excessive compression is applied repeatedly to the test phantom, cracks will form in the paddle, creating artifacts on the film.

The ***image receptor*** of mammography units also takes a great deal of abuse since the buckys are changed out frequently during the day. The fact is that the buckys are fairly heavy, weighing five to eight pounds, and are often dropped or mishandled during the course of the day. When a bucky is damaged, the most common complaint is the appearance of grid lines on the breast image, a sure sign of a loss of grid motion. With grid lines, the grid has either jumped off track after being dropped, or the grid motor has failed. In most cases, linear buckys can be repaired in the field, which helps reduce equipment downtime. However, because of the complex design of cellular grid systems, most manufacturers do not recommend repair in the field. Instead, service is handled on a repair/exchange program that ensures the quality of the repair and also helps reduce equipment downtime.

By far, the majority of mammography service calls are related to a ***loss of image quality;*** the most common of these calls deals specifically with ***density problems.*** Commonly, the technologist will notice a change in density on the phantom exposure that was taken on a particular morning and will call for service. Since high frequency x-ray generators are very reliable, the change in density is often a result of a variation in film processing or from a change in the batch of film being used. When the engineer responds (by phone) to the service call, he or she should try to rule these out as causes of the image density problem. The best way to isolate a density problem found on the phantom exposure is first to determine whether or not the mAs indication is correct–that it matches the value listed on the weekly QC charts. Second, the mAs indication should remain constant for every phantom exposure. If the mAs is consistent but the image density varies, the service engineer can confidently rule out the x-ray unit as the cause and can begin to look elsewhere for the problem. A change or fluctuating mAs indicates that the AEC circuitry requires calibration.

It is a fact that most service calls will be related to density issues. With so many variables that could cause a density problem, such as processor, film, dark room, and reading room variations, the engineer should first perform a thorough check of the entire mammography environment. It is suggested here that no changes be made until the reason for the density fluctuation has been found. Problems with film artifacts should be treated similarly. Here again, the environment must be completely ruled out first before the engineer begins to disassemble and troubleshoot the mammography unit.

Many times, service calls result from a failed radiation physicist's inspection. The radiation physicist performs routine inspections that thoroughly test every function of the mammographic unit. It is during these methodical tests that small errors in light field alignment or AEC density tracking are found. Though many of these errors do not critically affect the quality of the image, they nonetheless could cause increased exposure to the patient and, therefore, must be addressed. The engineer will respond to this type of service call by first requesting a copy of the physicist's report so that he or she will know exactly where to concentrate calibration adjustments.

Once service has been completed on the unit, the engineer before leaving should always "shoot" a phantom in order to verify proper operation of the entire system. Test films should be left at the site along with the service report to document calibrations.

THE FUTURE: DIGITAL MAMMOGRAPHY

Digital mammography offers many advantages over conventional film/screen mammography. First of all, digital mammography requires less radiation to obtain images. Also, the wider dynamic range of digital images essentially eliminates the need for ***retakes,*** which also reduces patient exposure. With digital imaging, the minor fluctuations in image densities are compensated for by altering the ***window*** and ***level*** settings at the control. Furthermore, the use of direct digital exposure eliminates all of the imaging problems caused by inconsistent film, defective intensifying screens, and darkroom and film processing variations. The high cost of purchasing and maintaining these auxiliary items is also eliminated.

Another advantage with digital mammography is that the images are available usually within seconds of the exposure, thus allowing the technologist to view images more quickly and to shorten examination time. As a result, many more patients can be screened in one day with a digital unit. Once the images are obtained, they can be archived and sent to remote locations for evaluation and storage. Finally, most digital units employ a ***CAD (computer aided detection)*** system to identify breast anomalies. With the CAD software, the computer analyzes the film, flagging possible carcinomas for the radiologist to examine at a later time. In effect, each film gets a "double reading" which greatly improves the detection of breast cancer.

Any disadvantages (i.e., reduction in image resolution and high cost) of earlier vintage digital units have been mostly eliminated as of this writing. Today, digital units can produce images that surpass the resolution achieved by the use of the normal film/screen systems. Also, high resolution monitors (10 lines pairs or greater) are now affordable and readily available. Many hospitals are currently using digital mammography systems, and soon most procedures will be performed on digital units.

Chapter XIII

ESTABLISHING AND MAINTAINING GOOD CUSTOMER RELATIONS

A book on x-ray servicing would not be complete without some mention of the relationship between the service engineer and the customer (i.e., the equipment owner or end user). After all, the ultimate goal of a service organization is customer satisfaction. In theory, if the customer is happy with the service being provided, that customer will continue using that particular service.

The responsibility for establishing and maintaining good customer relations lies on all members of a sales/service organization. Each individual must present the organization in a positive light and must be aware of his or her actions at all times. This is especially true for the sales and service staff members since they usually will have the most contact with the customer.

With a new equipment purchase, the customer will deal almost exclusively with a member of the sales staff of the equipment dealer. The salesperson will introduce the customer to the product, take the customer on site visits to observe the operation of the prospective equipment, negotiate the sale of the equipment, and meet with the customer as often as necessary to coordinate the installation of the equipment. Only occasionally, when his or her technical expertise is required, will it be necessary for the service engineer to meet with the customer.

However, once the sales agreement has been signed and the preliminary stages of the installation are well underway, the salesperson will begin to play a secondary role in dealing with the customer. At this time, the service engineer becomes the more visible representative of the equipment dealer, and must therefore be prepared to attend to other customer needs as they begin to surface. Furthermore, when the installation is completed, the x-ray service engineer may well become the *only* member of the x-ray equipment dealer (or service provider) who will regularly come into contact and interact with the customer. The responsibility for customer satisfaction will, therefore, depend solely on the service engineer.

Thus, it is clear that expertise in equipment servicing alone will not suffice when working in this field. The engineer must, in addition, possess the social skills required to establish and maintain a good relationship with the customer. Indeed, the electronic "wizard" who can fix any x-ray equipment on the planet cannot survive in the medical field if he or she cannot at the same time deal effectively with people.

Good customer relations benefits both the customer and the service provider as well. Usually, the satisfied customer will opt for future service with that provider which is financially beneficial to that provider. Conversely, a bad relationship resulting from the inappropriate actions of the service engineer can be very costly to the service provider.

For example, a service engineer may inadvertently so anger a staff member of the facility that the customer will respond by refusing to allow that particular individual to perform any further service on their equipment. Here, the service company will then normally comply with the customer's refusal and have to send another engineer to that site. Such an error, on the part of the engineer not only severely encumbers the service capabilities of the service organization, but also reflects poorly on the organization, as well as its personnel.

Furthermore, many service calls are not at all related to equipment failure. There are some customers who, unfortunately, may have a negative impression of a particular "brand" of equipment, regardless of its performance. This negative bias towards the equipment may have resulted from problems that the staff member had experienced at some time in the past at another institution, or it could be that a radiologist, for whatever reason, simply may not like the design of the equipment.

Whether preconceived or not, the engineer must attempt to understand why the customer feels a particular way about the equipment and then must remedy the situation. Sometimes, minor aesthetic changes made to the equipment can resolve this type of problem. Other solutions may involve modifications to the equipment (preapproved by the equipment manufacturer) to allow the equipment to operate to the customer's expectations. Here, the service engineer must be resourceful when resolving this type of customer problem. In essence, the service engineer is "fixing the customer."

It is an undeniable fact that good customer relation skills can lead to the success of a service organization. This is especially true, given the fact that most x-ray equipment manufactured today has very similar performance characteristics. Accordingly, the performance of the personnel of a service organization, especially the service engineer, can make the difference as to which equipment or service a customer will opt for in the future.

Because of the immense growth of personnel within this field, there are any number of service engineers available to "fix" x-ray equipment. In such a competitive field, it is vital for a service organization to find ways to maintain a competitive edge. Good customer skills can help maintain that edge. To achieve this goal, many x-ray dealers are now investing time and money into improving the customer relation skills of their engineers.

Although a complete course in customer relation skills may be a bit excessive, the engineer should be familiar with those factors that can effectively promote good customer relations. It has been the experience of the author that if the service engineer adheres to the common sense practices listed below, he or she will automatically promote and maintain good relations with the customer.

APPEARANCE

The service engineer should always dress professionally when servicing x-ray equipment. It is a simple fact that appearance makes a strong impression on the customer. A well-dressed service engineer projects a positive image for himself (herself) and the organization he or she represents. Proper dress for service engineers might well consist of a white or light colored shirt, a tie, and dark slacks, or a slight variation.

Because of the very nature of x-ray service, it is often difficult to maintain a clean appearance. During the course of a typical day, the engineer will be exposed to grease, oil, and other fluids that can permanently damage his or her clothes. Fortunately, inexpensive clothing is available today that is made of very durable, wrinkle free, wash and wear fabrics. The engineer can dress well without incurring a great expense for clothing and/or dry cleaning. Also, the tie can be tucked into the shirt, or removed, when actually performing service, thus avoiding the chance that it could get stained or, even worse, caught in a mechanical device.

ATTITUDE

The service engineer is in this business to solve problems–not to cause or enhance them. In most cases, the engineer is called to the site because the equipment has failed in some way. Consequently, he or she will regularly deal with people who are upset, frustrated, and even angry. If not prepared for these often tense situations, the engineer can easily become discouraged, or become negative.

When equipment "breaks down," it may completely disrupt the functions of the facility. This stressful environment requires a pleasant and calming demeanor on the part of the engineer to diffuse an otherwise explosive situation. By having a positive attitude when approaching a service call, the engineer will solve the problem much more easily and will leave behind a much happier customer.

Most causes for the strong emotions of the customer can, however, be easily resolved once the engineer begins to address the problem. If the engineer is positive in outlook, pleasant in manner, and greets the customer with a smile, he or she will achieve much better results. A positive attitude communicates that the problem is probably not so severe after all, and the engineer will be happy to take care of it. Being pleasant and amiable is a plus for that engineer, as well as the service provider.

RELIABILITY

The customer must be able to count on the service engineer for service. If the engineer schedules service with a customer, he or she should promptly arrive at the prearranged time. There is really no acceptable excuse for being late and making a customer wait. This communicates only that the customer is not important. The engineer should also be prompt when returning phone calls and should make sure that any promises or commitments are kept, as well.

COMPETENCE

The engineer must be confident in his ability to resolve x-ray failures and must demonstrate to the customer that he or she is capable of doing the work. Generally speaking, it will take only one or two service calls to gain the customer's confidence (or lack of it!). Oftentimes, a service engineer, while on a service call, will save the customer great expense by detecting (and fixing), in addition to the original failure, an annoying problem that has been intermittently occurring with the equipment. As a result, he or she will have won the confidence, as well as the goodwill of the customer.

If an engineer is unable to resolve an x-ray problem, the customer will be most appreciative if he or she is honest about it. No one person can repair every problem that may occur with x-ray equipment. If the engineer must call for assistance, the customer will simply note that the engineer did whatever it took to get the equipment operational.

COMMUNICATION

The service engineer must speak well when conveying information about a service problem to the customer, for the image he or she projects is that of the service organization. In addition, the engineer must clearly describe (in laymen's terms) a technical failure and then carefully explain how he or she repaired it. By doing so, the engineer shows that he or she respects the customer's intelligence and desires to keep him abreast of the situation.

To do this, the engineer must have the ability to convert highly technical details on equipment failures into "laymen's" terms that the customer can easily understand. It is often difficult to do this, given the complexity of x-ray equipment. If, however, the engineer can readily explain equipment failures (and their solutions) in relatively simple terms, the customer will feel more confident and comfortable with the service that is being performed.

An important part of good communication involves listening. The service engineer must have the ability to listen carefully to what the customer may be trying to say and should respond in the customer's best interest. "Best interest" might even entail offering the customer less expensive options or different solutions to a problem. The engineer who insists on repairing equipment, callously disregarding the cost to the customer, will soon be out of business.

It is very easy to underestimate the customer's ability to communicate and understand technical information. Engineers will often arrogantly ignore the customer's input on the service problem if they feel they already know what the problem is. Many customer relation problems result when the engineer repairs something that was not what the customer originally meant to be addressed, because that customer simply lacked the vocabulary or expertise needed to explain his or her problem thoroughly.

If the cause of the problem is found to be due to ***operator error,*** the engineer should in no way lay the blame on any individual for causing the failure. Instead, he or she should tactfully mention that a switch may have been depressed inadvertently, or if a device has been used in the wrong way, that its use may never have been properly demonstrated to the technician concerned. The engineer should then demonstrate the correct use of the equipment to eliminate similar service calls in the future. If the engineer decides that the person at fault will suffer consequences from a mistake (i.e., by being severely reprimanded by a superior), the engineer should then suggest, for example, that he or she has "discovered a loose connection and easily repaired it."

The author, here, is not advocating the practice of lying to the customer. However, as with every daily life interactions among people, a little "white lie" (as in the case above) may greatly aid in improving relations. The technician who made the mistake will more than likely recognize his or her error and appreciate the engineer's discretion. The engineer has now made at least one fast friend on that staff.

Clearly, it benefits absolutely no one to point a finger at an individual and will almost always result in a distrust of the engineer by many of the personnel at the hospital. If the service engineer acknowledges that human beings do make mistakes, and that x-ray technicians must often work with very sick or uncooperative patients, the temptation to make accusations will seldom arise. It should go without saying that the service engineer also makes, sometimes very foolish, mistakes!

ADDITIONAL SUGGESTIONS FOR IMPROVING CUSTOMER RELATIONS

- Listen carefully to the customer and address each concern individually. The best practice is to interact directly with either the chief technologist or the radiologist of that department.
- Try to offer alternative, less expensive solutions to problems with equipment.
- Do not "nickel and dime" a customer to death. Minor services should be provided at no cost to an established, loyal customer.
- Always remember to focus on patient and staff safety when dealing with any service issue.
- Make courtesy calls to customers even when equipment is functioning properly. Such a visit demonstrates a genuine concern for the customer and will help maintain good customer relations.
- Clean equipment thoroughly after performing service. Any marks left on the equipment reflects negatively on the service engineer.

Chapter XIV

THE "SEASONED" ENGINEER

To service medical instrumentation effectively, a service technician should possess skills in many areas of expertise. Service "techs" require strong mechanical skills, must be experienced in troubleshooting both analog and digital circuits, and should have knowledge of computer repair. In addition, all technicians should possess a good working knowledge of human anatomy and physiology, which will help them to better understand the design and application of medical equipment. Proficiency in all of these disciplines is essential simply because the equipment being serviced will be used to treat or diagnose patients, and their lives could depend on equipment accuracy.

Moreover, technical expertise is especially important for technicians servicing x-ray equipment where both dangerous voltages and lethal radiation levels are present. An incorrect calibration could cause serious injury to the patient or staff, not to mention the damage that could result to the sophisticated and very costly equipment. In fact, service personnel new to this field must follow an extensive training program before attempting to tackle the many challenges associated with servicing x-ray equipment.

It can take many years for service engineers to gain the expertise required to work effectively in this field. The engineer will have to perform countless calibrations and repairs on many different "brands" of equipment as he or she advances in the field. Only after experiencing each of the common types of failures occurring with radiographic equipment will he or she be able to diagnose and repair these failures efficiently.

The fact is that experienced engineers troubleshoot and repair equipment failures quickly and reliably, and when they leave the equipment site after a service call, they are usually confident that the unit will perform to manufacturer's standards. They have, in essence, become ***seasoned engineers.*** These experts have earned the respect of their supervisors, as well as their peers, and have become a valuable asset to their organization. Most customers note and appreciate the difference, too.

All service engineers would like to provide the best service possible, but realize that they must go through a learning curve which is a difficult and not always pleasant experience. This chapter will describe the various phases of an engineer's training and provide tips on how to get the most out of each phase while he or she is learning his or her trade. Next, servicing techniques are discussed which will help the engineer to develop skills so that he or she can perform repairs more reliably. The goal of this chapter is to help new engineers to attain the skill level of a "Senior Imaging Specialist" in the shortest period of time.

THE LEARNING PHASE

Anyone seriously interested in the field of x-ray service should first begin by learning as much as possible by reading everything he or she can on this subject, and through training in the classroom and laboratory. Once this ***education phase*** is completed, he or she must work in the field for some time serving an ***internship phase*** to gain the experience and confidence required for this job. It is during the internship that the engineer truly begins to learn the "ins" and "outs" of the trade. Usually working under the guidance of a senior engineer, the intern will put to test the practices learned in the classroom. This ideal setting takes much of the pressure off the intern so that the focus can remain on the equipment and servicing practices. As mistakes are made, the "mentor" can convert the error into a learning experience that will benefit the intern. The fact is that we often learn best from our mistakes: the embarrassment and frustration serve as a mental cue for future service calls. The key for newer service personnel is to take advantage of this internship phase, learning as much as possible and progressing through this necessary but awkward stage as quickly as possible.

At some point during this training, however, the intern will have to go it alone. There is only so much a person can learn from observing others, and the time comes when the engineer will perform service on his or her own. This important period of training is also the most difficult. Though it is vital that the engineer begin working independently, the concept of training while on the job can be very intimidating. After all, customers are paying for service and expect a "professional" to do the work. Engineers often comment on the frustration associated with these initial service calls because they sincerely want to satisfy the customer, yet they do not have the experience to perform their duties with confidence. The plain truth is that ***OJT (On-the-Job Training),*** commonly used in many specialized fields, is an effective means of training. During this training phase, it helps for the engineer to be mindful of the fact that he or she is not the only one who has gone through this phase–that everyone was a novice at one time!

Good or bad, the ***OJT phase*** is the way most engineers really develop their skills. As a rule, the most basic of service problems are assigned to the new engineer, and as experience is gained, the service calls will increase in complexity. As a representative of the service organization, the engineer should always strive to provide the best quality service. When confronted with an equipment failure not experienced before, the engineer should utilize all the resources that are available, such as consulting the operator and service manuals, calling coworkers, or even contacting the equipment manufacturer for assistance, if necessary. Many engineers ask members of the hospital staff to demonstrate equipment operation in an effort to learn about an equipment failure. The point is to take action and do whatever it takes to pin down the source of the problem and effect a repair. By taking an "active" approach, the engineer will quickly build the valuable experience needed for this job.

Every engineer must progress through each phase of the learning curve. But there are ways for an engineer to expedite this process and greatly shorten the time spent learning the trade. Included below are tips to aid the engineer to move quickly through the learning phases.

Understanding X-ray Theory

The first hurdle to overcome when beginning a career in the x-ray servicing field is to learn and understand fully the fundamentals of ***x-ray theory.*** Specifically, the engineer must know what x-ray radiation is, how it is produced, and how it interacts with matter to create the images used to diagnose medical problems. Possessing an in-depth knowledge of x-ray theory will greatly help when troubleshooting imaging problems. The engineer should focus on the effects that ***scat-***

ter radiation has on the final image. Also, with an awareness of the harmful effects of scatter radiation, he or she will be able to work confidently and ***safely*** around radiation producing equipment.

Learning Equipment Design

The newly hired service engineer should next concentrate on learning x-ray equipment design. He or she should know precisely how each device found in the x-ray room contributes to the process of creating and modifying the x-ray beam. The specific devices of the x-ray system, detailed in an earlier chapter, include the x-ray generator, the x-ray tube and tube support, the collimator, image receptor, wall receptor, and the radiographic table. Although designs may vary among the manufacturers, the function of each device remains the same.

For example, each type of x-ray generator has a distinctive output waveform, and there is a significant difference in the image formed from a single phase generator and one from a high frequency unit. Also, the effects of different x-ray tube types are striking. The engineer should know the consequences of using different target angles and filament sizes and should know the best application for each tube type.

Once x-ray radiation exits the x-ray tube port, anything lying directly in the path of the beam will alter its intensity and overall quality. The components of the collimator assembly, for instance, can attenuate, concentrate, soften, or totally block the x-ray beam. Many different types of x-ray grids are available that can drastically affect image quality. An engineer who knows the inventory of grid types can suggest alternatives to the customer, when asked. By doing so, he or she will communicate expertise and professionalism to that customer.

Learning Equipment Operation

Understanding equipment design and function is one thing, knowing how to operate equipment is another. All service professionals must take the time to ***learn equipment operation.*** In fact, a service engineer should know how to operate the equipment even before he or she can attempt a repair! X-ray equipment designs vary among manufacturers, and each design incorporates unique ways of accomplishing tasks such as programming technique factors, setting modes of operation, and selecting types of receptors. In addition, each unit has its own characteristic audible sounds and visual indications that identify the exposure sequence and indicate errors. Specifically, the engineer should note the distinctive sounds of the rotating anode of an x-ray tube, the duration of the prep cycle, and the sounds and indications of the x-ray generator during a normal exposure. These audible clues can be used when troubleshooting an equipment failure.

Initially, the best way to learn equipment operation is to ***observe the equipment while it is in use with patients during examinations.*** By doing so, the engineer will see the typical movements of the radiographic table and tube support during a routine x-ray procedure. Furthermore, he or she will see how an x-ray examination is performed, noting the sequence of events leading to the exposure.

For in-house engineers, observing patient exams is a fairly easy task. He or she can simply ask the chief technologist or the attending radiologist for permission to observe a few x-ray studies. In most cases the staff will be more than happy to show you their work. They may even point out a few things about the equipment that they like (or dislike). Almost certainly, the staff will appreciate your interest and will want to help you to better understand the specific nuances and quirks of the x-ray system.

Service engineers working for ***independent service organizations (ISOs)*** or ***original equipment manufacturers (OEMs)*** may have more difficulty getting into rooms for observation since they are not employed by the hospital and must account for their time. In reality, representatives of "third party" organizations are billing for their time and, in most cases, cannot afford the luxury of remaining at one site for very long. Before performing service, the field service engineers can ask the staff for permission to observe an examination "to learn clues about the failure." Also, the field engineer can follow up on a repair by observing an examination of a patient immediately after completing a repair to "ensure that the unit is working to specifications." This practice also communicates professionalism and helps promote customer relations.

During this observation time, the service engineer will see exactly how the equipment is used on patients, noting the sequence of events in common x-ray examinations and learning how different technologists perform the same examination. This information will prove very helpful when troubleshooting equipment failures and will greatly save time on future service calls.

The next step when learning equipment operation has to do with actual hands-on experience. The engineer should actually operate the equipment without the help of others until he or she feels comfortable with its operation. A good practice is to run through the process of setting up for an exam, programming the x-ray control, positioning the x-ray tube, and, ultimately, taking an exposure. In fact, all modes of operation should be tested. The engineer should expand this course of investigation, learning the operation of other equipment found in the radiology department. Specifically, the service person should learn the various nomenclatures used by different manufacturers, noting the variety of symbols and indicators that are used. At some point, the equipment operation will become second nature to the engineer and when called for service, he or she can speak intelligently about the equipment and can even demonstrate its operation with confidence.

Servicing New Equipment

When confronted with ***new or unfamiliar equipment,*** the engineer can begin using the general troubleshooting skills outlined earlier in this book. When questions arise, ask the staff for help. The hospital staff may realize that an engineer is unfamiliar with their equipment and will be more than glad to offer assistance so that they can get their unit "up and running." The first repair, no matter what it is, will be the most difficult one that an engineer will ever make. If an error was made in the troubleshooting process or in the repair itself, a return service call, or ***service recall,*** will remedy the situation. Again, the mistake will serve as a memory aid for future service calls.

The second service call on a similar failure will be much easier, but, still, the engineer may not be completely comfortable with the repair. The exact sequence of the repair procedure may not have been memorized, for example, or the engineer may not have remembered to take along the particular parts needed for the repair. The engineer may even have to place a phone call to the OEM for technical support, or ask a colleague for help. However, the third and fourth service calls on a similar equipment failure will be a "piece of cake." The engineer now has become an "old pro," and in the future, he or she will be the one giving technical advice to colleagues.

The Good Repair

When attempting to repair equipment, the engineer should strive to develop a specific and orderly routine that in time will become second nature to that engineer. Once perfected, that routine should be used with every repair. In this manner, the engineer will automatically create a system of inherent checks and balances that will ensure that equipment is repaired properly, thus reducing the chance of errors.

First, when disassembling components, the engineer should arrange the parts on a surface (e.g., table or floor) that is free from clutter so that there will be no chance of misplacing or losing the parts. As each part is removed from the assembly, it should placed in order on the table maintaining the exact spatial relationship so that it will be much easier to reassemble. Using a clock as illustration, the top cover to an assembly is placed at twelve o'clock, bottom cover at six o'clock, the left cover at nine o'clock, and so on. All screws, washers, spacers, and gaskets are placed on the table maintaining their exact relationship. By matching the exact spatial orientation of the parts, the engineer can reassemble them more confidently and more quickly. This practice works well with parts that are very similar in appearance.

Another good habit is to place cover screws or other similar hardware in a tray or small container so that they will not get knocked around or lost. If a work surface is not available, a good practice for the engineer is to label the parts to avoid any confusion. The few extra steps taken here will reduce the chance of error during reassembly.

When the failing component is replaced, it should be tested thoroughly. The engineer should watch for an obvious change in equipment performance with the new components in place: there is really no gray area here. In fact, if there is any question at all in the engineer's mind that the original symptom may still exist (even intermittently), the engineer should begin looking elsewhere for the problem. With x-ray equipment, there could be multiple sources for a given problem, and the engineer must investigate all of the possible causes of the failure. A rule of thumb is that if not completely confident with the repair, the engineer can expect to return a second time to repair the same problem.

Once the unit is completely reassembled (i.e., all covers in place), the engineer should perform a functional test of the equipment. As stated in the troubleshooting chapter (Chapter IX), subassemblies and wire harnesses are packed closely together within the equipment cabinets, a drawback to the compact design of x-ray equipment. It is common in x-ray servicing to have equipment develop symptoms of a failure once the equipment covers are back in place–wires get crimped and connectors can become dislodged. It is therefore recommended that the engineer test all modes of operation before leaving the site. An even better practice is to invite a staff member to test the equipment and verify its operation.

Record Keeping

A practice that will guarantee quick advancement for new engineers relates to documentation. Good record keeping is an organizational practice that has proven absolutely necessary in many professions. Most seasoned engineers will take ***detailed notes*** on specific equipment failures that he or she has experienced and will document the information gleaned from other valuable sources such as from coworkers, technical articles and books, and from course instructors. In fact, the best engineers will maintain a service log containing a history of specific service calls and their associated resolution. These engineers will also document unique calibration procedures that are difficult to perform and will record any unusual test waveforms that he or she has encountered.

The simple fact is that the mind cannot possibly retain every detail of the failures occurring with so many types of equipment. Furthermore, when just starting out, there is simply too much information for the engineer to digest. The engineer who documents the various equipment problems that he or she encounters over the years will be way ahead of the game. The notes can be studied or used as a reference on future service calls.

The Extra Effort

When out on a service call, the main goal for the engineer is to get the equipment back up and running as quickly as possible. While performing the repair, however, he or she may notice other minor problems with equipment in the radiographic room. The technologist or staff member may even point out an annoying "quirk" with the equipment. These minor problems are often "tolerated" for a period of time, but the staff will often not call service, sometimes just to save money. Here is a perfect opportunity for the senior engineer to take action–and "shine."

The engineer can perform small acts that will make a huge difference in the staff's perception of the service performance. Lubricating a noisy mechanical assembly, for example, or replacing missing hardware or burned out indicator lamps are actions that can be performed with little effort. Also, the engineer can clean or adjust switches so that they "snap" into position when operated. By performing this additional service, not only is the failure corrected (i.e., the original reason for the service call) but the engineer has eliminated an annoyance that had been bothering the radiology staff for some time. Most top notch engineers perform these small tasks without mentioning it to the staff. The overall impression left is that the equipment just seems to be working much better after that particular engineer has been there. The extra effort taken by the service professional will truly pay off in the long run.

Furthermore, the additional service performed may even have prevented a future service call. True, every service organization has to make money to stay in business–and major service calls are expected from time to time–however, it is the little nagging calls that often create a negative view of the equipment, or the service engineer, or both. Also, it is not always a good thing to have an engineer present at the site on a regular basis. The customer may wonder why the engineer is always working on the equipment. Frequent visits often reflect poorly on the engineer as well as on the equipment which may have been purchased from that service engineer's company.

In this author's view, it is more beneficial to make minor repairs while on a service call, rather than having to return frequently to the site. There is a another benefit, as well. Happy customers will tell others of their experience and thus, promote more business for the engineer and his or her service organization.

SUMMARY

It takes time to learn the skills needed to repair radiographic equipment, and given enough time, many students can learn to become good engineers. However, engineers who want to advance quickly and excel in this field must demonstrate initiative early on and take an active approach to servicing. If the engineer adheres to the guidelines outlined above, he or she can remain on the fast track, and soon will become a ***seasoned engineer.***

APPENDICES

APPENDIX A

MEASURED kVp	MINIMUM REQUIRED HVL (in millimeters of aluminum)
30	.3
40	.5
50	1.2
60	1.3
70	1.5
80	2.3
90	2.5
100	2.7
110	3.0
120	3.2
130	3.5

Figure 43. HVL Chart.

APPENDIX B

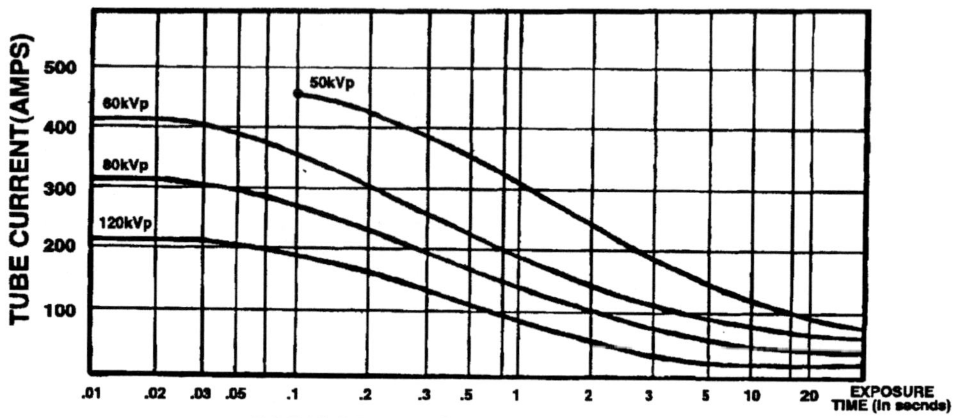

RADIOGRAPHIC EXPOSURE CHART

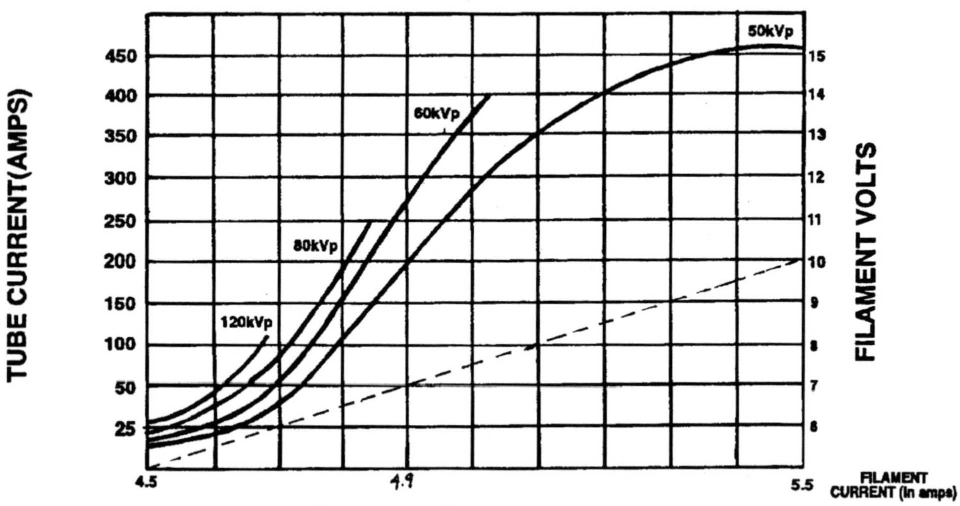

FILAMENT EMISSION CHART

Figure 44. Sample x-ray tube rating charts. (Top): At a setting of 80 kVp, 300 mA, the maximum allowable exposure time is 0.05 seconds. Also, notice that at 50 kVp an emission limit is reached. Therefore, the maximum mA obtainable for this tube is 460 mA at 50 kVp. (Bottom): Note the emission limit at 50 kVp is also graphically represented on the filament emission chart. Also, at 50 kVp we see that 4.9 amps of filament current is required to achieve 200 mA. The dotted line in the chart represents the voltage/current relationship of the filament wire. Specifically, if 10 VAC is applied to the filament wire, the actual measure filament current would be 5.5 amps.

APPENDIX C

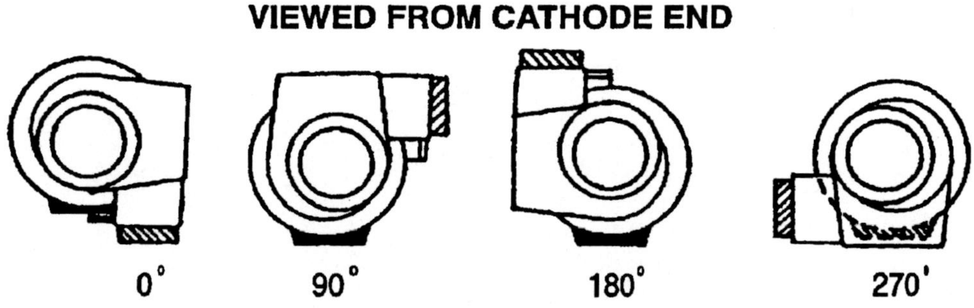

Figure 45. Horn angle designation for x-ray tube housing.

APPENDIX D

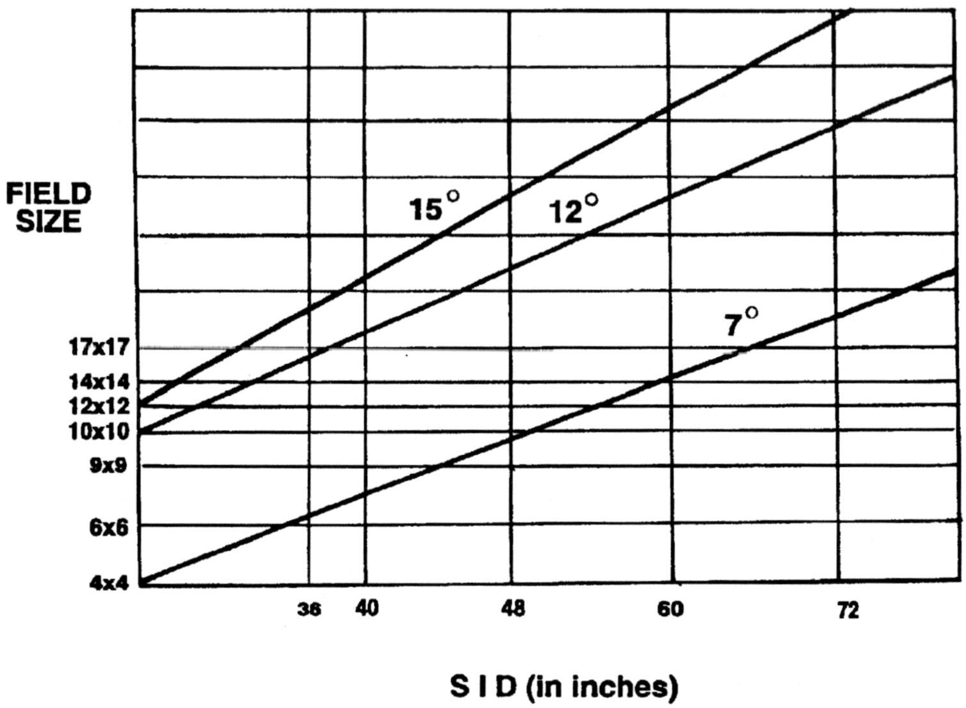

Figure 46. X-ray field coverage for various target angles. Note that a minimum of 40 inches SID is required to achieve 17 inch film coverage with a 12 degree target.

APPENDIX E

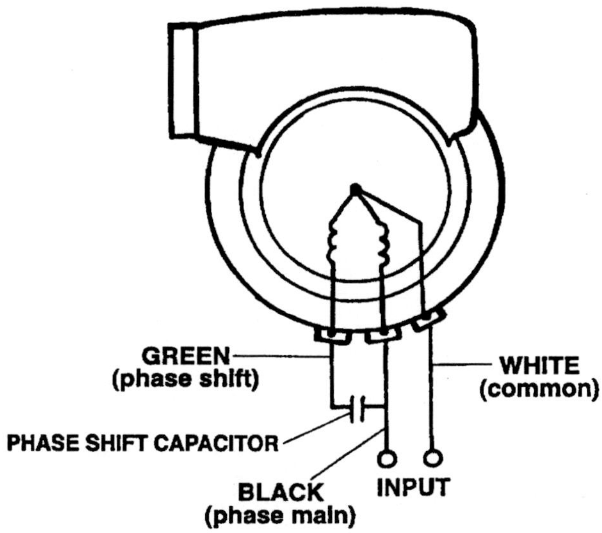

STATOR TYPE	IMPEDANCES (Typical) Black - White	Green - White
"R" Type (Standard)	16	50
"Q" Type (Low Impedance)	9	16
"E" Type (Balanced)	23	23

Figure 47. Common x-ray tube stator winding impedances.

APPENDIX F

TOOLS AND TEST EQUIPMENT

Certain specialized tools and test equipment are needed to service x-ray equipment properly. Besides the standard tool kit used for servicing general electronic devices, additional equipment is required for moving and installing radiographic equipment. For x-ray installation and servicing, the following tools and special equipment are needed.

X-ray Service Tool List

1. A Complete Electrician's Tool Kit
2. 3/8″ Electric Drill (light duty)
3. 1/2″ Electric Drill (heavy duty)
4. Hammer Drill (for concrete)
5. Handtruck and dolly (moving equipment)
6. High Jack (moving)
7. Johnson Bar (moving)
8. Rollers (moving)
9. Come Along (moving)
10. Electrician's Fish Tape (wire pulling)
11. Socket Set (standard and metric)
12. Open End Wrench Set (standard and metric)
13. 4 foot Spirit Level or precision level (alignment and mounting)
14. Plumb Bob (alignment and mounting)
15. Crow Bar (moving)
16. Hacksaw
17. Vacuum Cleaner (installation and PM)

It may not be practical for a service provider such as in-house groups to own all of the above equipment because of cost or space limitation; however, the engineer must have access to these tools in order to provide complete x-ray service. Smaller x-ray companies can rent the moving equipment as needed, and in-house groups can borrow equipment from the plant operations or maintenance departments.

In order to properly service x-ray equipment, specialized *test equipment* is also required. The service engineer must be able to measure high voltage and radiation output during calibrations. To properly diagnose problems, the engineer must view the x-ray waveform on an oscilloscope. To check the condition of the x-ray tube and imaging chain, special test phantoms are required. Test equipment needed for x-ray service is listed below. All of the test equipment is mandatory for servicing radiographic equipment.

Test Equipment

1. Digital Volt Meter (DVM)
2. Ammeter (AC clamp type)
3. Digital Storage Oscilloscope (DSO)
4. High Voltage Bleeder or Dynalyzer
5. MAs Meter
6. Dosimeter ("r" output)
7. Non-invasive kVp Meter
8. Light Meter or Photometer (collimator light output)
9. Flat Field Phantom (AEC)
10. Line Pair Phantom (system resolution)
11. Pie Mesh Phantom (system resolution)
12. Lead Star Test Phantom (x-ray tube focus)
13. Densitometer (AEC calibration)
14. Sensitometer
15. Laptop Computer (generator programming and troubleshooting)
16. Vibrating Reed Tachometer (anode rotational speed)

APPENDIX G

TOMOGRAPHIC ANGLES vs. SLICE THICKNESS

Angle	*Thickness*
5°	11.0 mm
10°	6.0 mm
20°	3.0 mm
30°	2.0 mm
40°	1.4 mm
50°	1.1 mm

GLOSSARY

ACR–The American College of Radiology. Establishes standards and practices for mammography.

AL-1100–The type of aluminum alloy used to attenuate the x-ray beams. It is used in beam limiting devices and for testing purposes.

ARTIFACT–Unwanted, additional information located on the x-ray film (or imaging medium) that is caused by a defect in the imaging chain, a film processing problem, or by improper handling of the x-ray film. Film artifact is a serious problem that can cause misdiagnoses.

BUCKY–A device that contains a reciprocating grid assembly positioned directly in front of the film. A slot is also provided in the bucky assembly for the AEC detector. Originally called the Potter-Bucky, named after the two inventors: Dr. Hollis E. Potter and Dr. Gustav Bucky.

CASSETTE TRAY–The device that holds the film cassette in position during the exposure. The cassette tray may have cassette size sensing circuitry that is used in conjunction with automatic collimators.

COLLIMATOR–A lead-lined beam limiting device with adjustable shutters that allow the operator to vary the size of the x-ray field. All collimators provide a field lamp for field size indication.

CONTRAST–The difference in densities between two adjacent areas on the image.

CONE–A lead-lined device that mechanically attaches to the x-ray tube used to limit the x-ray beam. All cones have a fixed field size and do not provide a light localizer.

CRAZING–A phenomenon that occurs in x-ray tubes caused by internal arcing. This arcing causes heat expansion inside the tube that results in fractures or cracks of the glass insert. Crazing can be observed by looking directly into the port.

DIAPHRAGMS–A fixed-aperture beam limiting device that attaches directly to the x-ray tube housing. Diaphragms are usually removable.

DENSITOMETER–A calibrated device used for measuring film density. Densitometers are used for film processor quality control and during the AEC calibration by service engineers.

DENSITY–The amount of blackening of film measured by a densitometer.

FALLING LOAD–The mA waveform produced by a special type of x-ray generator. A falling load generator will initially produce the maximum mA allowable for the particular x-ray tube at the beginning of the exposure and then will drastically reduces the mA after a preprogrammed interval. The resulting radiation waveform will have a flat portion followed by a negative slope.

FORCED COMMUTATION (or forced extinction)–A method of quickly turning "off" an SCR contactor by actively removing the source voltage. The SCR is usually forced "off" by a second SCR.

GREY–The international measurement of absorbed dose. 1 Gy = 1 joule of radiation energy absorbed per kilogram of tissue. 0.01 Gy = 1 RAD.

GRID–A device used in radiography that effectively reduces scatter radiation, thus improving image quality. The grid is placed directly in front of the image receptor.

HEEL EFFECT–A phenomenon that occurs in all x-ray tubes caused by the target angle and the anode thickness. Defined as the variation in exposure rate with the angle of emission of the radiation from the focus. Radiation emanates in all directions from the anode. The radiation that travels toward the anode end of the x-ray field must pass through a small portion of the anode and is, consequently, attenuated. This results a slight decrease in radiation intensity and is clearly visible on the film.

HVL–Half Value Layer. A measurement unit that represents the quality of the x-ray beam. It is the amount of AL-1100 required to reduce the R output of the x-ray tube by 1/2. Federal guidelines for minimum requirements for HVL are provided so that patients will not receive excessive soft radiation (see Appendix A).

HIGH VOLTAGE DIVIDER–A test instrument used to visualize and measure the actual kVp waveform. This divider is essentially a metal tank which contains an internal resistor divider network surrounded by insulating oil. It is connected in series with the x-ray tube and used in conjunction with an oscilloscope.

H.U.–Heat Units. A measure of the power (heat) capacity of an x-ray tube. The amount of heat an x-ray tube can safely withstand is a function of the target size and tube housing type. One heat unit is equal to the product of the kilovoltage, tube current, and the exposure time (kVp * mA * TIME) for single phase generators. Because of the improved efficiency of a three-phase generator, the heat applied to the x-ray tube is increased by a factor of 1.35, for 6 pulse, and 1.41 for a 12 pulse system.

ILLUMINATOR–Also called a view box. A device used by radiologists for viewing radiographs.

INTENSIFYING SCREEN–The material that lines the inside of film cassettes. Intensifying screens convert x-ray radiation to light radiation which then exposes the film.

IMAGE RECEPTOR–A device that transforms x-ray radiation that has passed through the patient into a visible image (i.e., after further processing as in film). In radiography, the receptor is the film and the film cassette.

INHERENT FILTRATION–The amount of filtration that exists in an x-ray tube alone. The tube naturally has filtration caused by the combination of the glass insert, insulating oil, and tube housing port. The amount of inherent filtration should be an equivalent of .5mm of aluminum.

K-ABSORPTION EDGE–The threshold or limit, specific to the type of filter material used, above which radiation will be absorbed.

MAMMOGRAPHY–A specialized field of radiography developed for the screening and diagnosis of breast cancer.

MQSA–Mammography Quality Standards Act, established in 1994.

OID–Object to image distance. The distance from the object being x-rayed (i.e., phantom or patient) to the film.

PHANTOM–A test device that simulates patient tissue and is used by field service engineers and radiation physicists for image evaluation. Phantoms are made of a variety of materials including acrylic, copper, aluminum, special fluids, and lead. Phantoms are mainly used to test image resolution and contrast; however, the service engineer also uses a flat field phantom to measure optical densities of film.

RAD (Radiation Absorbed Dose)–A unit of measurement for the amount of radiation absorbed by the patient. It is a measurement of the total entrance dose delivered in tissue. One RAD represents the absorption of 100 ergs of energy per gram of tissue.

REM (Radiation Equivalent for Man)–A measurement of radiation absorbed dose that is multiplied by a quality factor dependent on the type of radiation being used. For diagnostic x-ray radiation, 1 RAD = 1 REM.

RESOLUTION–A measurement of an imaging system's ability to produce sharpness. Also refers to image sharpness or definition.

ROENTGEN (R)–The unit of measurement of radiation (free in air). One R is defined as the amount of radiation required to produce one electrostatic unit (ESU) of charge in 1 cc of air under standard conditions of pressure, temperature, and humidity.

ROTOR–Used in reference to the rotating anode of an x-ray tube. The rotor is the moving part of induction motor.

SAFELIGHT–The light used in a darkroom that contains a low wattage light bulb and a special filter that will not expose x-ray film. The correct filter must be matched to the film being used.

SCATTER RADIATION–A characteristic of radiation caused by the interaction with matter. Scatter radiation adversely affects image quality by reducing image contrast.

SCOUT FILM–A test film taken prior to a special procedures examination to ensure proper positioning technique.

SELF-COMMUTATION–A method of turning "off" an SCR by allowing the SCR to turn "off" with the naturally occurring zero cross-over points of the AC line.

SENSITOMETER–A device used for processor quality control. It exposes film to produce a calibrated step wedge of increasing densities. Measurements are made with the resulting "strip" for film base fog, speed, and contrast indexes.

SHARPNESS–The abruptness of change between two adjacent densities (i.e., abruptness of the boundary). A fast change between two different densities that are adjacent to each other will produce a sharper image than a gradual change.

SID–Source to image distance. The distance from the focal spot of the x-ray tube to the film plane.

SIEVERT–The international unit for absorbed dose normalized for the type of radiation used. 1 Sv = 100 REM.

SPACE CHARGE–The cloud of electrons that forms around the cathode when the filament is heated.

SSD–Source to skin distance. The distance from the x-ray tube to the patient's skin.

STATOR–Used in reference to the stator windings of a rotating anode in the x-ray tube. The stator windings create the magnetic field which, through induction, causes the rotor to turn.

TOMOGRAPHY–A radiographic technique in which a slice of anatomy is imaged (also termed linear tomography).

TRUNION RINGS–A special tube mounting bracket that holds the x-ray tube in place and allows tube rotation, if desired.

VIBRATING REED TACHOMETER–A test instrument used by x-ray service engineers that measures the rotational speed of the rotating anode in an x-ray tube. This device senses the rotation of the anode by the frequency of vibrations felt through the tube housing.

ZONOGRAPHY–A technique used in tomography where a thick section, or zone, is imaged on film by using a very small tomographic angle.

BIBLIOGRAPHY

Cullinan, John E.: *Illustrated Guide to X-RAY TECHNICS.* Philadelphia: J.B. Lippincott, 1972.

The Fundamentals of Radiography (12th ed.). Rochester, NY: Eastman Kodak Company, 1980.

Gofman, John W. & Egan O'Connor. *X-rays: Health Effects of Common Exams.* San Francisco: Sierra Club Books, 1985.

Grigg, E.R.N.: *The Trail of Invisible Light.* Springfield, IL: Charles C Thomas, 1965.

Hendee, William R.: *Medical Radiation Physics* (2nd ed.). Chicago: Year Book Medical Publishers, 1979.

Heywang-Kobrunner, Sylvia H. et al.: *Diagnostic Breast Imaging: Mammography, Sonography, Magnetic Resonance Imaging, and Interventional Procedures* (2nd ed.). Trans. Peter F. Winter. Stuttgart, New York: Thieme, 2001.

Jacobi, Charles A.: *X-ray Technology.* Saint Louis, MO: C.V. Mosby, 1964.

Nitske, Robert W.: *The Life of Wilhelm Conrad Roentgen, Discoverer of the X-Ray.* Tucson, AZ: University of Arizona Press, 1971.

Selman, Joseph: *The Fundamentals of X-ray and Radium Physics* (6th ed.). Springfield, IL: Charles C Thomas, 1977.

Thompson, Edward P.: *Roentgen Rays and the Phenomena of the Anode and Cathode.* New York: D. Van Nostrand, 1896.

INDEX